Physiology and management of mangroves

Tasks for vegetation science 9

Series Editor

HELMUT LIETH

University of Osnabrück, F.R.G.

Physiology and management of mangroves

edited by

H.J. TEAS

1984 **DR W. JUNK PUBLISHERS**
a member of the KLUWER ACADEMIC PUBLISHERS GROUP
THE HAGUE / BOSTON / LANCASTER

Distributors

for the United States and Canada: Kluwer Boston, Inc., 190 Old Derby Street, Hingham, MA 02043, USA
for all other countries: Kluwer Academic Publishers Group, Distribution Center, P.O.Box 322, 3300 AH Dordrecht, The Netherlands

Library of Congress Cataloging in Publication Data

Main entry under title:

Physiology and management of mangroves.

   (Tasks for vegetation science ; v. 9)
   Papers presented at the Second International
Symposium on the Biology and Managment of Mangroves,
held at Port Moresby, Papua New Guinea in 1980.
   1. Mangrove plants--Congresses. 2. Coastal zone
management--Congresses. 3. Coastal zone management--
Tropics--Congresses. I. Teas, Howard J. (Howard Jones),
1920-   . II. International Symposium on the Biology
and Management of Mangroves (2nd : 1980 : Port Moresby,
Papua New Guinea) III. Series.
QK938.M27P49 1983   583.42   83-19932
ISBN 90-6193-949-6

Contents

CONTENTS OF

Biology and ecology of mangroves

H.J. TEAS (ed.)

Published as Volume 8 of the series Tasks for Vegetation Science

Professor G.V. Joshi (August 19, 1925–July 24, 1982)

IN MEMORIAM

GOVIND VISHNU JOSHI
1925–1982

G.V. Joshi received his Doctorate from the University of Bombay. In 1967, after 17 years on the faculty of Wilson College in Bombay, he became Head of Botany at Shivaji University at Kolhapur. During 1959–61 he was a National Academy of Sciences Postdoctoral Fellow at the University of Southern California.

Although most of us knew Joshi for his outstanding work on mangroves, his field was basically plant physiology. He worked principally on photosynthesis and mineral metabolism. His investigations involved marine algae, mangroves and other halophytes as well as a few crop plants. Joshi published more than 50 papers, mostly on photosynthesis, enzymes of photosynthesis, and photosynthesis and relationships of mangroves to salinity. He presented papers at both the First International Symposium on the Biology and Management of Mangroves in Hawaii in 1974 and at the Second Symposium at Port Moresby, Papua New Guinea in 1980.

Professor Joshi served as President of the Indian Society of Plant Physiologists and on a variety of UN and other agency and government committees and study groups. He was editor-in-chief of the Indian Journal of Plant Physiology. In 1981 Professor Joshi was awarded the J.J. Chinoy Gold Medal as 'The Best Plant Physiologist in India' by the Indian Society of Plant Physiologists. He was clearly an outstanding scientist by international standards as well.

Professor Joshi was a scholar. In addition, he was a kindly and patient man with a sense of humor and a good teacher. Fifteen students finished their doctorates under his direction.

Professor Joshi's contributions were great. He is missed by his many friends and colleagues.

HOWARD J. TEAS

CHAPTER 1

Photosynthesis and photorespiration in mangroves

G.V. JOSHI*, SHUBHANGI SONTAKKE, LEELA BHOSALE and A.P. WAGHMODE

Botany Department, Shivaji University, Kolhapur 416004, India

Abstract. It has been shown in our earlier work that in many mangroves aspartate is synthesized within a period of less than five seconds and to avoid intense light and high temperatures of midday, the stomata remain wide open in the early morning up to 10–11 a.m. In the present investigation this has been confirmed in *Lumnitzera* and *Ceriops*. A further probe into the intricate process of photosynthesis in *Rhizophora mucronata* Lamk, *Rhizophora apiculata* Bl., *Ceriops tagal* (Perr) C.B. Rob, *Bruguiera gymnorrhiza* (L) Lamk, *Avicennia officinalis* L., *Avicennia marina* (Forsk.) Vierh. *Aegiceras corniculatum* (L) Blanco, *Lumnitzera racemosa* (Willd) and *Acanthus ilicifolius* L. by supplying uniformly labelled aspartate and studying enzymes of carbon fixation has also been attempted. In these mangroves the activity of PEP Case is considerably higher than that of RuBP Case and justifies synthesis of aspartate as the initial product. The activity of pyruvate-Pi-dikinase can be detected in all these mangroves which provides continuous supply of PEP for carbon assimilation. The decarboxylating enzyme in the form of NAD-malic enzyme is also detected. CO_2 compensation point in *Lumnitzera* and *Ceriops* are low. The pulse chase experiment with $^{14}CO_2$ and labelled aspartate, the photosynthetic enzyme studies and photorespiration in *Lumnitzera* and *Ceriops* suggest operation of a modified C_4 path in the mangroves. It is argued that such a pathway is a physiological response to ecological conditions which impose stress on the ecosystem.

Introduction

The role played by the mangrove vegetation in the ecosystems of the world is being increasingly understood today. According to Pannier (1979) mangroves represent a typical ecosystem found along many tropical coasts and estuaries. As a nutrient filter and synthesizer of organic matter, mangroves create a living buffer between land and sea. The SCOR/UNESCO panel (1976) has stated 'Scientists are agreed that mangrove areas, far from being waste lands, are highly productive and comparable to good agriculture land'. According to Teas (1977) mangroves are valuable for shoreline protection and stabilization as well as habitat for wild life and sources of photosynthetic productivity. In spite of this, there is little information on the basic process of photosynthesis in mangroves. This information is all the more necessary because these plants have a mechanism which will be useful in silviculture and agriculture with saline water. Teas (1979) has tried to use this basic information on mangroves while discussing silviculture with saline water. Similarly the information on photosynthesis of mangroves

* (Paper prepared for Second International Symposium on biology, and management of mangroves and tropical shallow water communities, Port Moresby, Papua-New Guinea, July 1980)

will be of help while understanding agriculture with saline water. Hence there is a need for increasing research activities in the various aspects of photosynthesis in mangroves.

Historical

As early as 1891 Schimper observed that the leaves of the mangroves show a special structure that corresponds to xerophyte leaves. He correctly visualized that the problem involved here was that of metabolism. The first comprehensive review which appeared on halophytes was that of Adriani (1958), however, although he cited no reference to carbon metabolism in mangrove species, he reviewed valuable information on the importance of salt concentration for carbon assimilation. Adriani quoted Fromageot (1924) as having shown that optimal carbonic acid assimilation in *Ulva lactuca* occurred at the salt concentration of sea water and that respiration was also dependent on salt concentration. Montfort and Brandrup (1927) reported similar observation for *Zostera marina* and for *Ruppia rosellata*. They gave a very broad ecological amplitude of 0–12% salinity for carbon assimilation. If there was an increase in optimal salt concentration there was also a depression in carbon assimilation which might be due to chlorosis (Lessage, 1891; Keller, 1925; Arnold, 1955).

Carbon assimilation in succulents is unique in that they take up carbon dioxide at night, a process which is known as Crassulacean Acid Metabolism (CAM)*. Halophytes, and mangroves were investigated by Bennet-Clark (1933) and Warick (1960) and it was found that they do not have the machinery for CO_2 uptake at night.

Recent work on mangrove photosynthesis

In the recent years the first comprehensive review on mangroves is that of Macnae (1968). He has stated that eventhough mangroves can grow under nonsaline conditions in an experimental environment not many mangrove species are found to be successful competitors in upland or freshwater environments. According to Walsh (1974) little is known about the production of organic matter in mangrove swamps. He stated that the first detailed study of photosynthesis, respiration, biomass and export of organic matter was that of Golley *et al.* (1962).

The work these investigators carried out was not on carbon assimilation mechanisms but their rate by different plant parts. They found that although the net photosynthetic rate for sun leaves was much greater than that for shade leaves, the bulk of carbon assimilation was not carried out by the sun leaves at the top of the canopy but by shade leaves, a finding which was confirmed by Miller (1972). Chapman (1976) noted that relatively little work had been carried out in the field of carbon assimilation.

Photosynthesis in succulent or non-mangrove halophytes

Considerably more information is available on photosynthesis in succulent halophytes than on mangroves. According to Chapman (1976) the mangroves are essentially halophytes and their physiology should be comparable to that of other halophytes like *Salicornia, Arthrocnemum* and *Suaeda*. A good deal of work has been carried out on these plants by the group of Shomer-Ilan and Waisel (Shomer-Ilan & Waisel, 1973; Shomer-Ilan *et al.*, 1975, 1979; Beer *et al.*, 1976). These workers

Abbreviations

CAM	–	crassulacean acid metabolism
DTT	–	dithiothreitol
EDTA	–	ethylenediaminetetracetate
MES	–	2(N-morpholino)ethanesulfonic acid
NAD-ME	–	nicotinamide adenine dinucleotide dependent malic enzyme
NADP-ME	–	nicotinamide adenine dinucleotide phosphate dependent malic enzyme
PCK	–	phosphoenolpyruvate carboxykinase
PEP Case	–	phosphoenolpyruvate carboxylase
PVP	–	polyvinylpyrrolidone
PCK	–	phosphoenolpyruvate carboxykinase
RuBP Case	–	ribulose-1,5 bisphosphate carboxylase

have shown that these succulent halophytes are C_4 plants without Kranz anatomy. They have depicted a path of carbon in the salt marsh plants where two types of chloroplasts exist in two different layers. They have also shown that there was a mechanism of aspartate decarboxylation in mitochondria and that CO_2 was refixed by the Calvin cycle. They demonstrated that in *Atriplex spongiosa* there was a C_4 path with malate and aspartate as the initial products. They also showed that there was no photorespiration and that rates of CO_2 fixation in this species were three times greater than in a C_3 type, *A. hastata*. The C_4 succulent had low RuBP Case and glycolate oxidase and very high PEP Case. According to the view of Chapman (1976) on the similarity of the metabolism in nonmangrove halophytes and mangroves there should be a modified C_4 path of photosynthesis in mangroves, like non-mangrove halophytes.

Work on photosynthesis in mangroves

The first report of CO_2 fixation in saline plants by Rho (1959), and further elaborated by Joshi *et al.*, (1962), conclusively demonstrated the preponderance of amino acid synthesis over that of organic acid. They also showed inhibition of malic dehydrogenase and stimulation of transaminase in a seawater medium. This was further confirmed by Webb and Burley (1965) in salt marsh halophytes, which included *Salicornia* sp. According to them this amino acid shift was due either to the effect of NaCl on the carboxylation reaction or to pool sizes of organic acids in individual species.

The first report on photosynthetic products in mangroves appeared in 1974 (Joshi *et al.*) on *Aegiceras corniculatum* (L.) Blanco. Aspartate was reported to be the initial product. This finding had special significance, because a comprehensive treatise on photosynthesis and photorespiration (Hatch *et al.*, 1971) had appeared and the C_4 path had been firmly established as the path of tropical plants under stress. Subsequently, Long and Woolhouse (1978) showed that the C_4 path of photosynthesis was not restricted to tropical species but was also found in *Spartina townsendii*, a Graminaceous

plant occuring in temperate intertidal mudflats. Moore *et al.*, (1973) studied net photosynthesis, dark respiration, and transpiration of *Rhizophora mangle, Avicennia germinans* and *Laguncularia racemosa* and found that there were seasonal variations and that the highest values occurred in summer. According to Golley *et al.* (1962) light saturation occurs at 5,000 f.c. (53,800 lux) which is about 50% of incident light in the tropics. There are also reports of low photorespiration in mangroves. Lewis and Naidoo (1970) studied transpiration rates in *Avicennia marina* and found that they were high at mid morning and low in the afternoon. The low rates are attributed to the incipient wilting resulted from excessive transpiration. This suggested that the stomata were wide open only in early mid mornings when photosynthesis must take place.

A comprehensive paper on photosynthetic carbon metabolism in mangroves was presented (Joshi *et al.*, 1975) at the first International Symposium on the Biology and Management of Mangroves. India has more than twelve mangrove species which occur on the east and west coasts of the country. These investigators selected *Rhizophora mucronata* Lamk., *Ceriops tagal* (Perr.), *Bruguiera gymnorrhiza* Lamk., *Avicennia officinalis* Linn., *Aegiceras corniculatum* (L.) Blanco, *Lumnitzera racemosa* (Willd) and *Acanthus ilicifolius* Linn. for a study of the products of photosynthesis by the pulse chase method after supplying $^{14}CO_2$. They found that aspartate and alanine appeared as the major initial products which were subsequently utilized for synthesis of sugars and other metabolic products. Preliminary experiments indicated dominance of PEP Carboxylase (PEP Case) over that of RuBP Carboxylase (RuBP Case).

The present paper summarizes our work that has been carried out since the First International Symposium on Mangroves and attempts to interpret the results in the light of the findings of other investigators on saline photosynthesis. Besides the above mentioned mangroves, *Rhizophora apiculata* Bl., *Avicennia marina* (Forsk) Vierh. and *Lumnitzera racemosa* (Willd.) have been added. In these mangroves aspartate and alanine also appear to be the major initial products of photosynthesis. In order to confirm the metabolic role of aspartate in light carbon assimilation, uniformly labelled aspartate was supplied to *Aegiceras corniculatum, Ceriops tagal,* and *Lumnitzera racemosa* and metabolic products were studied after one hour. The results confirm the important role of aspartate in carbon assimilation which can be further elaborated in view of C_4 photosynthesis. PEP Case, RuBP Case and pyruvate, Pi-dikinase were studied and methods for their assay have been standardized. In addition to suggesting the dominance of PEP Case over RuBP Case, the presence of pyruvate, pi-dikinase is thought to supply PEP for initial carboxylation of the C_4 path in mangroves. The decarboxylation reactions appear to be controlled by NAD-ME. The process of photorespiration in *Ceriops tagal* and *Lumnitzera racemosa* have been studied by measuring CO_2 compensation point, feeding of glycolate 1–^{14}C and glycine 1–^{14}C, measuring activity of glycolate oxidase and glycine decarboxylase and rate of glycolate synthesis. The results show that photorespiration is lower in the mangroves than in sunflower (C_3) and that mangroves tend to resemble the C_4 type sugarcane. However it may be mentioned that the works on stomatal physiology and transpiration control as well as on energy transfer, chloroplast morphology, enzyme studies in photosynthesis and its localization are for the most part lacking and they will be needed to understand photosynthesis in the mangroves so that their productivity and that of other salt marsh plants can be better understood.

Materials and methods

The mangrove field sites are located about 140 km from Kolhapur in the Ratnagiri district. Twigs and saplings of the mangrove species were collected and brought to the laboratory. Leaves of *Rhizophora mucronata* Lamk., *R. apiculata* Bl., *Bruguiera gymnorrhiza* Lamk., *Aegiceras corniculatum* L. Blanco., *Avicennia marina* Forsk., *A. officinalis* (L.), *Acanthus ilicifolius* (Linn.), *Sonneratia alba* J.Sm in Rees and *S. apetala* Buch. Ham., *Exoecaria agallocha* (Linn.), *Ceriops tagal* Perr., and *Lumnitzera racemosa* Willd, along with

the mangrove grass, *Aeluropus lagopoides* L., and a seagrass *Halophila beccarii* Aschers., were used for the investigation.

Leaf discs of the mangroves (25 mm^2, cut by a sharp punching machine) were floated in Tris-HCl buffer (pH 7.8) and supplied with NaH^{14}CO$_3$ or ^{14}CO$_2$ for a period of 2–5 seconds or for periods up to one hour. The details of pulse chase study methodology have been described in an earlier communication (Joshi, 1976). Similarly uniformly labelled aspartate was supplied in light and dark to the leaf discs of *Aegiceras corniculatum, Ceriops tagal, Lumnitzera racemosa* and *Aeluropus lagopoides* for periods ranging from 5 seconds to 1 hour.

Enzymes

Healthy leaves of *Rhizophora mucronata* Lamk., *R. apiculata* Bl., *Bruguiera gymnorrhiza* Lamk., *Avicennia marina* Forsk., *A. officinalis* L. and *Acanthus ilicifolius* L. were collected. Laminae were cleaned and 1 g was ground and extracted in a medium containing 50 mM Tris HCl (pH 7.5), 2 mM MgCl$_2$, 1 mM EDTA, 40 mM mercaptoethanol, 1.5% polyvinyl pyrrolidone(PVP) and 10 mM sodium metabisulphite. The extract was centrifuged at 10,000 g for 10 min. The supernatant served as the crude source of enzyme. The temperature maintained during extraction was 0 to 5° C for PEP Case and RuBP Case and 22° C for pyruvate, phosphate di-kinase.

PEP Case and RuBP Case were estimated by the procedure described by Das and Raghavendra (1976). 10 μl of NaH^{14}CO$_3$ was directly added in the assay mixture. The rate of activity fixed into the acid soluble product was determined with a 2π Proportional Counting System.

Pyruvate, phosphate di-kinase

The assay procedure is based on that of Hatch and Slack (1969) which is a coupled method that requires PEP Case and amino transferase enzymes. Because a crude extract is used, these enzymes are assumed to be present endogenously and hence are not added. When the ingredients were added in micromolar quantities, the values obtained were very low and hence various concentrations of major ingredients described by Hatch and Slack (1969) such as pyruvate (2.2 μM, 1 mM, 2 mM, 10 mM), ATP (1.1 μM, 2 mM, 10 mM) and K_2HPO_4 (0.37 μM, 2 mM) were tried. The best results were obtained with the reaction mixture (1 ml) containing 50 mM tris HCl (pH 7.5), 1 mM pyruvate, 10 mM $MgCl_2$, 0.1 mM EDTA, 2 mM K_2HPO_4, 10 mM mercaptoethanol, 10 mM ATP, 1.3 mM sodium glutamate and 0.1 ml extract. To this mixture 10 μl $NaH^{14}CO_3$ (activity 50 mCi/mmole) was added. The reaction was stopped by 0.5 ml of 20% TCA after 2 min incubation at 25 ($\pm$ 1 C). Blanks were run simultaneously in which pyruvate or ATP or enzyme was absent.

Unit

For PEP Case, RuBP Case and pyruvate, phosphate di-kinase, the unit is defined in terms of 1 μ mole of $^{14}CO_2$ fixed under standard assay conditions.

Protein estimation

Protein was estimated according to the method of Lowry *et al.*, (1951). Since polyphenols interfere with the method, the extract was treated with five volumes of acetone, then the protein precipitate obtained was dissolved in an adequate amount of the buffer.

Malic enzyme

This was estimated by the method of Kluge and Osmond (1972). One g of leaf discs was extracted with 200 mM bicine, 10 mM $MgCl_2$, 170 mM 2-mercaptoethanol and 16 g/l PVP at pH 8.3. The assay contained in a final volume of 1 ml, 0.5 ml of enzyme extract, 50 μ moles MES, 50 μ moles bicine, 10 μ moles $MgCl_2$, 10 μ moles mercaptoethanol, 0.37 μ moles of NAD/NADP and 6 μ moles malate at pH 7.2.

CO_2 compensation point

CO_2 compensation point was determined by the method of Das and Raghavendra (1973).

Leaf anatomy

The leaves of above listed mangroves were hand sectioned, stained with I_2KI and examined immediately. The specimens were examined for the presence of bundle sheath chloroplasts and Kranz anatomy.

Stomatal behavior

Stomatal behaviour was studied by the method of Stoddard (1965). The width of stomatal aperture was measured under a calibrated microscope on nail polish films.

Glycolate oxidase

Glycolate oxidase was assayed spectrophotometrically by the method of Hess and Tolbert (1967) in which the increased absorbence due to formation of phenylhydrazone was measured at 324 nM. The reaction mixture of 3.0 ml contained 2.4 ml of 0.1 M phosphate buffer, 0.1 ml of 0.1 M cystein hydrochloride, 0.1 ml of 0.1 M phenyl hydrazine hydrochloride, 0.3 ml of 0.1 M sodium glycolate and 0.1 ml enzyme.

Rate of glycolate synthesis

Circular leaf discs of 25 mm^2 weighing 1 g were placed in Petri plates in distilled water. After stabilizing in natural sunlight the distilled water was replaced by 10 mM sodium metabisulfite. After 1 hr incubation in the same stabilized condition the leaf material was immediately killed with 4 N HCl and after elution with N HCl from alumina column the glycolate was estimated according to the method of Calkins (1943). The rate of glycolate synthesis was determined in terms of μg fwt^{-1}hr^{-1}.

Glycine decarboxylase

Activity of glycine decarboxylase was determined by trapping $^{14}CO_2$ released from glycine-1-^{14}C by the method of Woo and Osmond (1976). The reaction mixture in 1.5 ml consists of 0.4 M sucrose, 0.1 M glycylglycine buffer pH 7.5, 5 mM K_2HPO_4 and 20 mM (-1-^{14}C) glycine.

L/D ratio

Photorespiratory ratio was determined by trapping released $^{14}CO_2$ from glycolate-1-^{14}C and glycine-1-^{14}C in light and dark, in 20% KOH by the method of Kisaki and Tolbert (1970).

Glycolate-1-^{14}C and glycine-1-^{14}C were obtained from BARC, Bombay, India, had sp. act. 1.9 mCi/m mole and 9.3 mCi/m mole respectively. The total chlorophylls were estimated by the method of Arnon (1949) and proteins by the method of Lowry et al., (1951).

Results and discussion

The results on carbon assimilation in mangroves are to be viewed in light of the earliest observation of xeric features of mangroves by Schimper (1891); the salt dependence of marine plants for carbon assimilation reported by Fromageot (1924); more recent observation of Chapman (1976) that the process of carbon assimilation in mangroves can be similar to that of saline succulents like Suaeda; and to the recent thinking on C_4-pathway which gives advantages to the plants growing under stress. The process of light carbon assimilation will depend on the stomatal opening and duration, chlorophylls, carbon assimilating enzymes, and the machinery to utilize immediately the products of photosynthesis.

There are few reports on the stomatal physiology of mangroves. The works of Lewis and Naidoo (1970) indicate that stomats are likely to be open in mid-mornings and close completely in the afternoon. Joshi et al. (1975) reported that maximum stomatal openings for several species were at 10 a.m. Two more mangroves were investigated here,

namely Ceriops tagal and Lumnitzera racemosa and it was found that the maximum stomatal opening is at 11 a.m. It was found that stomata are nearly closed in midday and afternoon and completely closed at night. For the leaves with erect position the stomata are only on the lower side and these are also covered by hairs. All these facts clearly indicate that little time and little exposure is available for photosynthesis. There is no mechanism such as occurs in succulents for CO_2 storage at night and there are not two photosynthetic bursts namely one in the mornings and another in the late noon when temperatures are low. Das and Shantakumari (1977) have indicated that in C_3 plants stomata are open up to 10 a.m. while they can remain fully open till midday in C_4 plants because of their capacity for photosynthesis at higher temperatures. It is not possible to arrive at any definite conclusion on the basis of this criterion about C_3 or C_4 nature of the mangrove but the only thing that can be said is that mangroves must assimilate CO_2 at a faster rate. As per the reports of Golley et al., (1962) the highest rate of carbon assimilation occurs in the sun leaves.

The chlorophyll contents of the mangroves investigated are given in Table 1. Chlorophyll contents of leaves from a few nonmangrove species are also listed for comparison. It appears that man-

Table 1. Chlorophyll contents of the leaves of mangroves and a few known C_3 and C_4 plants

Name of species	Total chlorophylls (mg/100 g fresh wt.)
Rhizophora mucronata	58
Bruguiera gymnorhiza	113
Ceriops tagal	105
Sonneratia alba	88
Aegiceras corniculatum	135
Avicennia alba	81
Avicennia officinalis	105
Acanthus ilicifolius	66
Excoecaria agallocha	168
Lumnitzera racemosa	71
Salvadora persica	52
Sugarcane (C_4)	250
Aeluropus lagopoides (C_4)	220
Parthenium hysterophorus (C_3)	180

groves have less chlorophyll than other plants. The average value for total chlorophylls in mangroves was usually less than 150 mg/100 g while it typically was 150 or higher for other plants. This may be due to 'dilution' by water storage tissue, or more fiber content in the leaves. This low chlorophyll content of mangrove leaves suggests that CO_2 assimilation must follow a faster and more efficient path.

Role of aspartate in mangrove photosynthesis

Joshi et al. (1975) demonstrated that aspartate and alanine are the initial photosynthetic products in A. corniculatum, A. ilicifolius, S. alba, R. mucronata and E. agallocha. We have extended those studies to two more mangroves, Ceriops tagal and Lumnitzera racemosa, and the results of short term $^{14}CO_2$ fixation in the seven mangroves are shown in Table 2. These results further confirm that PGA is not the initial product of photosynthesis in mangroves, but rather that aspartate (and alanine) is. It is possible that aspartate originates from CO_2 being fixed on OAA by PEP Case and the keto acid is immediately converted to aspartate because the pathway to malate synthesis is blocked due to inhibition of malic dehydrogenase in the saline media (Joshi et al., 1962; Joshi, 1976).

In order to obtain a conclusive proof of this reaction the leaf discs of A. officinalis, C. tagal and S. alba were exposed to $NaH^{14}CO_3$ in the light for a period of less than two seconds and the initial products trapped as phenylhydrazine derivatives, the latter were separated by buffered paper chromatography (Das and Rathnam, 1974) and the incorporated radioactivity measured. The results are shown in Table 3. It is clear that OAA had appreciable label and this was passed to aspartate which also has considerable radioactivity. These products are typical of C_4 plants which are of aspartate type. The enzyme which forms aspartate from OAA, aspartate amino transferase, is being sought in the mangroves. It may be mentioned that in the saline grass, Aeluropus litoralis, (Shomer-Ilan and Waisel, 1973) aspartate appeared as the initial product. They have also shown salt dependant activity of PEP Case in the salt tolerant grass. In our laboratory Waghmode and Joshi (1979) have shown that in the mangrove grass, Aeluropus lagopoides, aspartate is the initial product of light CO_2 fixation and it is metabolically utilized. Similarly, Das and Raghavendra (1977) have reported malate and aspartate synthesis in salty sandbinder, Spinfex squarrosus. These studies suggest that aspartate is the initial product of photosynthesis in saline plants. It has been argued (Hatch, 1975) that alanine may be derived from PGA and that aspartate synthesis may represent a simple dark reaction of CO_2 assimilation. The detection of radioactive OAA and metabolic utilization of aspartate clearly suggest that this is a main reaction of photosynthesis in the mangroves.

The pulse chase experiments revealed that with the lapse of time the activity in aspartate decreases while that of other compounds, inclusive of sugars and sugar phosphates increases.

In order to further confirm metabolic utilization

Table 2. Initial products of photosynthesis in mangroves. Activity expressed as the percentage of total activity counted on chromatogram.

Name of the Plant	Aspartate	Alanine	Malate	Citrate	Succinate	Glutamate	PEP + PGA	Origin	Other
Aegiceras corniculatum*	36	57	–	–	–	–	7	–	–
Acanthus ilicifolius*	40	60	–	–	–	–	–	–	–
Sonneratia alba*	49	32	5	–	–	–	14	–	–
Rhizophora mucronata*	60	32	–	–	–	–	17	–	–
Ceriops tagal**	32	33	–	–	–	14	–	16	5
Lumnitzera racemosa**	16	58	–	–	–	2	–	21	3
Excoecaria agallocha**	23	49	11	11	8	–	–	–	–

*   From Joshi et al. (1975)
** Present investigation, values for 10 sec.

Table 3. Activity in phenylhydrazine derivative of OAA after a period of less than two seconds of photosynthesis of NaH^{14}CO$_3$ by mangroves.

Species	Aspartate	Phenylhydrazine derivative of OAA	Origin	Other spots of phenylhydrazine derivatives
Avicennia officinalis	49	22	–	29
Ceriops tagal	54	10	–	26
Sonneratia alba	32	32	23	13

Activity counted in aspartate, phenylhydrazine derivative of OAA and origin are expressed as percentage distribution of total activity on chromatogram.

of aspartate, leaf discs of *Ageceras corniculatum, Ceriops tagal, Lumnitzera racemosa* and *Aeluropus lagopoides* were floated in the buffer and fed with uniformly labelled aspartate in light and dark. In light, labelled aspartate was utilized rapidly and sugars were formed while in dark the activity remained in aspartate only. These results further confirm a metabolic role of aspartate in photosynthesis. Imai *et al.,* (1975) have extensively studied rates of sugar formation from radioactive compounds in C$_3$ and C$_4$ plants. They also studied the rate of sugar formation from aspartate-^{14}C-(U) and alanine-1-^{14}C under various light intensities in three C$_4$ plants. Sugar formation from aspartate-^{14}C(U) was much greater in C$_4$ plants than in C$_3$ ones. According to them this is due to lack of enzymatic machinery in C$_3$ plants. The metabolic utilization of aspartate-^{14}C-(U) in the mangrove leaf discs shows that there are enzymes for its utilization.

Carboxylating enzymes

In the last Symposium (Joshi *et al.* 1975) reported preliminary results which indicated that of the two carboxylating enzymes, PEP Case was more active than RuBP Case. A difficulty in working with enzyme preparations from mangroves is the presence of salts and polyphenols which inactivate the enzyme during extraction. This problem was overcome by addition of EDTA, mercaptoethanol, PVP and sodium metabisulphite. The basic method employed is that of Das and Raghavendra (1976). Results of enzyme measurements are shown in Table 4. These results confirm our observation that

in the mangroves PEP Case is more active than RuBP Case, a feature characteristic of C$_4$ plants. The role of PEP Case in saline succulents has been explained by Shomer-Ilan *et al.,* (1979). They had earlier showed the salt dependent nature of PEP Case in *Aeluropus litoralis* (1973). It is well known now that PEP Case is a better and faster fixer of CO$_2$ and hence plays an important role in plants with stress. In mangroves, stomata are open for a short time, the temperatures are high and water is a problem and hence in these xeromorphs PEP Case can be a natural choice for carbon assimilation. The activity of PEP Case in the mangroves ranged from 6.7 to 37.8 μ moles/mg of protein. The ratio of PEP Case/RuBP Case varied from 2 to about 9 (Joshi *et al.,* 1980). Raghavendra and Das (1978) have also obtained similar results for *Setaria italica, Pennisetum typhoides* and *Amaranthus paniculatus*, all C$_4$ species. The ratio in these plants varies from 3 to 6. These results clearly indicate that in mangroves, the C$_4$ pathway of photosynthesis is operative. Even though our values reported in this paper utilize a modified extraction medium, they appear to be low when we compare them with typical C$_4$ glycophytes. There is still a scope for improvement of procedure in the isolation and assay of enzymes in the mangroves. We suggest that in mangroves PEP Case may be localized in the cytosol or on the chloroplast envelope while RuBP Case is in the chloroplasts.

Pyruvate, phosphate di-kinase

For the continuous activity of carbon assimilation by PEP Case there must be a continued supply of

PEP. This is provided in C_4 plants by the enzyme, pyruvate, phosphate di-kinase. This enzyme was first detected in grasses like maize, sugarcane and sorghum (Hatch and Slack, 1968). Subsequently this enzyme was detected in many C_4 plants (Hatch and Slack, 1975; Raghavendra and Das, 1978). The enzyme is localized in mesophyll tissue of NADP-malic enzyme type, NAD-malic enzyme type and PCK type of C_4 plants (Ku and Edwards, 1975). According to Hatch *et al.* (1975) the major role of mesophyll chloroplasts is to convert pyruvate to PEP via this enzyme. This enzyme is not present in C_3 plants and its detection in the mangroves is all the more important.

The enzyme pyruvate, phosphate di-kinase is a difficult enzyme to extract in an active form even in glycophytes: it has complex and varying requirements for activation of extracts (Hatch and Slack, 1969). The presence of salt and polyphenols in high concentration add to the difficulties. After repeated attempts a method has been standardized by us (Joshi *et al.*, 1980). The difficulties are overcome by adding 1.5% PVP and 10 mM sodium metabisulphite in the extraction medium. When PVP was added to give more than 1.5% concentration, PEP Case is inhibited to about 66% while RuBP Case was slightly activated. Therefore, 1.5% was chosen as the ideal concentration. PVP acts as an adsorbent for polyphenols and sodium metabisulphite is used as an antioxidant. Use of DTT or 2-mercaptoethanol alone cannot give satisfactory extraction.

Pyruvate, phosphate di-kinase is a cold sensitive enzyme and for this enzyme the extraction was carried out at room temperature $22 \pm 1°C$. When the activity was estimated in the same extract as for PEP Case and RuBP Case 0–5°C negligible amount of activities was obtained. This observation is in accordance with the results obtained by Hatch and Slack (1969).

The activity of pyruvate, phosphate di-kinase in the six mangrove species is shown in Table 4. It ranged from 2.5 to 14.0 μ moles/mg of protein. The ratio of PEP Case to pyruvate, phosphate di-kinase ranged from 1.5 to 4.8. Hatch (1975) showed that in *Portulaca oleracea* the activity of PEP Case was five times more than that of pyruvate, phosphate di-kinase. Similar observations have been made by Raghavendra and Das (1978) in the above mentioned species. The activities of this key enzyme in the mangrove species are on similar lines of those of C_4 species. The ratio may not be high in all the mangroves but the trend is on that line. The presence of this key enzyme of C_4 photosynthesis is highly significant. Attempts to demonstrate light dependence of this enzyme are in progress.

Decarboxylation reactions

In C_4 photosynthesis decarboxylation reactions are highly significant. These reactions provide basic CO_2 for the Calvin cycle which provides the carbon skeleton for synthesis of important compounds in the plant. In 'aspartate former' type of C_4 plants the decarboxylation of aspartate takes place either by NAD-ME or PCK. In order to study this, NAD-ME was studied. It was found that this enzyme is very active in light and its activity is neglegible in the dark (Fig. 1). The activity is also higher than that of NADP-ME.

Table 4. Photosynthetic enzymes in mangroves (Activity expressed as μ moles of product mg protein^{-1} 5 min^{-1}).

Plant	PEP Case	RuBP Case	Pyruvate, phosphate di-kinase	PEP Case/ RuBP Case	PEP Case/ Pyruvate, pi dikinase
Avicennia marina	14.4	7.2	4.1	2.0	3.51
A. officinalis	12.1	2.6	2.5	4.65	4.84
Rhizophora mucronata	17.0	2.4	4.5	7.08	3.78
R. apiculata	20.2	2.3	14.0	8.78	1.44
Bruguiera gymnorrhiza	6.7	2.1	8.5	3.19	0.79
Acanthus ilicifolius	37.8	15.0	6.5	2.52	1.50

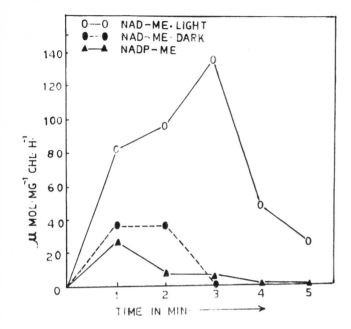

Fig. 1. NAD-ME & NADP-ME in *Rhizophora mucronata*

Light stimulated CO_2 release is known as photorespiration. The rate of photorespiration is much lower in C_4 plants than C_3 ones due to efficient refixation of any CO_2 released by active PEP Case. Photorespiration is due to the formation of glycolic acid in the chloroplast and its subsequent oxidative metabolism to give CO_2 (Halliwell, 1978). Glycolate produced by chloroplasts is first oxidized to glyoxylate in the leaf peroxisomes. Most of glyoxylate is transaminated to glycine. The rest of the glyoxylate produced reacts non-enzymatically with H_2O_2 generated by glycolate oxidase to yield CO_2 and formate. Another example of CO_2 release is during conversion of glycine to serine in the leaf mitochandria. In the present investigation an attempt is made to study the subsequent steps of photorespiration (Table 5).

Activity of glycolate oxidase is higher in C_3 plants and lower in C_4 plants and this can be taken as one of the criteria in distinguishing C_3 and C_4 plants (Andreeva *et al.*, 1975); Kundu *et al.*, 1975; Osmond & Harris, 1971; Rehfeld *et al.*, 1979). Glycolate oxidase of the halophytes along with sugarcane and sunflower is studied by Waghmode and Joshi (1979). Their results showed that the ratio of activity of glycolate oxidase in typical C_3 and C_4 types was consistant with previous results and that halophytes are more like C_4 type than C_3 because the activity of glycolate oxidase is low and nearer to that of sugarcane than sunflower.

Shomer-Ilan *et al.*, (1979) have shown high NAD-ME activity in chlorenchyma I layer of *Suaeda monoica,* a succulent halophyte, where decarboxylation of aspartate takes place and CO_2 is fixed by the Calvin cycle. This is a mitochondrial enzyme. This other enzyme of C_4 photosynthesis is also present in the mangroves. The detection of pyruvate, pi-dikinase and NAD-ME is of significance and demonstrates C_4 photosynthesis in mangroves.

Table 5. Release of $^{14}CO_2$ from glycolate-1-^{14}C and glycine-1-^{14}C as a substrate, rate of glycolate synthesis and activities of glycolate oxidase and glycine decarboxylase.

	$^{14}CO_2$ evolved from glycolate-1-^{14}C cpm/g fresh wt.			Glycolate oxidase n mol $min^{-1}mg^{-1}$ (protein) (Waghmode and Joshi) 1979	Rate of glycolate synthesis $\mu g/hr/g$ fresh wt.	$^{14}CO_2$ evolved from glycine-1-^{14}C cpm/g fresh wt.			Glycine decarboxylase cpm/g fresh wt.
	Light $\times 10^3$	Dark $\times 10^3$	L/D			Light $\times 10^3$	Dark $\times 10^3$	L/D	$\times 10^3$
Ceriops tagal	0.8	1.5	0.57	21	350	7.2	14.5	0.49	33.3
Lumnitzera racemosa	1.4	1.1	1.22	12	20	6.8	6.3	1.08	56.7
Sugarcane (C_4)	3.4	22.4	0.17	20	20	5.7	7.8	0.75	10.7
Sunflower (C_3)	18.8	9.7	1.94	73	ND	13.5	2.5	5.47	18.7

ND – Not determined.

Waghmode and Joshi (1979) have further shown a 90 and 93 percent inhibition of glycolate oxidase of *Ceriops* and *Lumnitzera* by 5 mM sodium metabisulfite. The rate of glycolate synthesis when investigated by using 10 mM sodium metabisulfite showed values typical for C_4 in *Lumnitzera* but higher ones for *Ceriops*.

There are different opinions regarding the substrate for photorespiration. Kisaki and Tolbert (1970) have shown glycine as the better substrate while Zelitch (1972) was of opinion that glycolate was a better substrate. Oliver (1979) studied the mechanism of decarboxylation of glycine and glycolate by isolated soybean cells and found that about half of the CO_2 released from glycolate occurred during conversion of glycine to serine, the remainder was released in the direct oxidation of glyoxylate by H_2O_2. Screening experiments with mangroves using glycolate and glycine showed that glycine was the better substrate for photorespiration, which was further confirmed by the activity of glycine decarboxylase (Table 5). The L/D ratio using glycolate-1-^{14}C and glycine-1-^{14}C revealed that rate of photorespiration was more typical of C_4 than C_3 plants (Table 5). Kennedy (1976) has compared photorespiration rates in tissue cultures of a C_4 plant, *Portulaca oleracea* to *Streptanthus torturosus*, a C_3 plant and has shown that C_4 plant tissue cultures have 1/2 to 1/3 the photorespiratory rate of the C_3 plant. Results of the present investigation are parallel. In *Lumnitzera* utilization of glycolate-1-^{14}C was 0.4% and 2.2% in light and dark respectively (Table 6). Rapid metabolism of glycolate-1-^{14}C is indicative of an efficient photorespiratory path and a feature of C_3 path (Berlyn *et al.*, 1978). Thus by several criteria photorespiration in mangroves is low, resembling that of C_4 plants.

Recently Krause *et al.*, (1978) have shown that photorespiration is essential for the plant cell to protect photosynthetic apparatus from damage. The photorespiration of mangroves although low, may have such a metabolic role.

Model for photosynthesis in mangroves

The data accumulated so far are not adequate to

Table 6. Utilization of glycolate-1-^{14}C in 1 hour light and dark by *Lumnitzera racemosa.*

Compound	1 h light	1 h dark
Glycolate*	99.6	97.8
Glyoxylate	0.3	1.0
Glycine-serine	0.1	–
Citrate	–	0.5
Unknown	–	0.4
Origin	–	0.3

(Values expressed as percentage ^{14}C on paper chromatogram)
* Exogenously supplied

construct a comprehensive model for photosynthesis in mangroves. This is unlike the case of succulent halophytes where Shomer-Ilan *et al.* (1979) have developed such a model. The known photosynthetic enzymatic reactions that have been demonstrated in mangroves are shown in Fig. 2. It appears likely that in mangroves atmospheric CO_2 enters through sunken stomata which are open for a very short time during the day. CO_2 is assimilated in the spongy mesophyll cells by PEP Case which is either in the cytoplasm or on the chloroplast envelope. The OAA formed is rapidly converted to aspartate but not to malate because malic dehydrogenase is strongly inhibited by the NaCl within the cells. The synthesised aspartate can serve as a pool for synthesis of other amino acids and a major part of it probably enters specialised mitochondria where it is decarboxylated by NAD-ME and the released CO_2 is re-fixed in the chloroplasts by the Calvin cycle. The location of these two types of chloroplasts is not certain. The pyruvate which enters the chloroplasts is converted to PEP by pyruvate-pi-dikinase which has been shown to be present in the chloroplasts. However, there are several uncertainties in this model. The role of bundle sheath chloroplasts, if they are present in the mangrove leaves is still to be elucidated. The localization and separation of metabolic reactions in the carbon cycle are still to be demonstrated. The coordination of C_4 products with that of C_3 cycle is still to be shown. There is little information on stomatal physiology and chloroplast morphology of the mangroves. What is known is

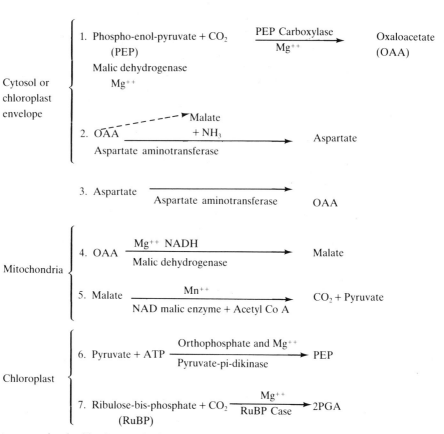

Fig. 2. Important enzymes involved in photosynthesis in mangroves

minimal and what is yet to be understood is enormous.

Acknowledgements

The investigation forms a part of research project sponsored by the Department of Science and Technology of Government of India on 'Photosynthesis and Marine Environment' (Project No.7(15)/76 SERC). The enzyme portion of the studies was possible because of a grant USA from P.L.480 Project on 'Crop Productivity of Kolhapur Region' (Project No.TN-ARS-102) provided through the US Department of Agriculture. The authors are grateful to these agencies for their support.

Literature cited

Adriani, M.J. 1958. Halophyten In Encyclopaedia of Plant Physiology Vol.IV 709–736 Springer-Verlag Berlin.

Andreeva, T.F., T.A. Avdeeva and S.Yu Stepanenko, 1975. Effect of nitrogen nutrition on glycolate oxidase activity in bean and corn plants. Fiziol. Rast. 22:553–557.

Arnold, A. 1955. Die Bedeutung der Chlorionen fur die pflanze. Jena: Gustav Fischer.

Arnon, D.I. 1949. Copper enzymes in isolated chloroplasts. Polyphenoloxidase in *Beta vulgaris*. Plant Physiol. 24:1–15.

Beer, S.A. Shomer Ilan and Waisel, 1976. Salt stimulated phosphoenol pyruvate carbxylase in *Cakile maritima*. Physiol. Plant. 34:293–295.

Bennett-Clark, T.A. 1933. The role of organic acids in plant metabolism. New Phytol. 32:37–71, 128–161, 197–230.

Berlyn, M.B., I. Zelitch and P.D. Beaudette. 1978. Photosynthetic characterisitics of photoautotrophically grown tobacco callus cells. Plant Physiol. 61:606–610.

Calkins, V.P. 1943. Microdetermination of Glycolic and oxalic acids. Industrial and Engineering Chemistry, 15:762–763.

Chapman, V.J. 1976. Mangrove vegetation. J. Cramer. In der A.R. Gantner Verlag Kommanditgesellschaft FL. 9490 Vadus.

Das, V.S.R. and A.S. Raghavendra. 1973. A screening of the dicot weed flora for the occurance of C_4 pathway of photosynthesis. Proc. Indian Acad. Sci. 77:93–100.

Das, V.S.R. and A.S. Raghvendra, 1976. C_4 photosynthesis and a unique type of Kranz anatomy in *Glossocordia boswallaea*. Proc. Indian Acad. Sci. 84 B(1):12–19.

Das, V.S.R. and A.S. Raghvendra, 1977. Kranz leaf anatomy and C₄ dicarboxylic acid pathway of photosynthesis in *Spinifex squarrosus* L. Indian J. exp. Biol. 15:645–648.

Das, V.S.R. and C.K.M. Rathnam. 1974. Mechanisms of regulation of plant growth. The Royal Society of New Zealand. Wellington Bull, 12:223–228.

Das, V.S.R. and M. Santakumari, 1977. Stomatal characteristics of some dicotyledonous plants in relation to the C₄ and C₃ pathways of photosynthesis. Pl. Cell Physiol. 18:935–937.

Fromageot, C. 1924. Sur les relations entre l'état physicochimique et le Fonctionnement du protoplasma: Photosynthese et respiration. Bull. Soc. Chim. biol. Paris 6:69.

Golley, F.B., H.T. Odum and R.F. Wilson, 1962. The structure and mechanism of a Puerto Rican red mangrove Forest in May. Ecology, 43:9–16.

Golley, P.M., F.B. Golley Eds. 1972. Tropical ecology with an emphasis on organic productivity. Athens. Ga: Int. Soc. Trap. Ecol., Int. Assoc. Ecol. Ind. Nat. Sci. Acad. 418 pp.

Hatch, M.D. and C.R. Slack, 1969. Studies on the mechanism of activation and inactivation of pyruvate phosphate dikinase. A possible role for the enzyme in C₄ dicarboxylic acid pathway of photosynthesis. Biochem J. 112:549–558.

Hatch, M.D. 1975. C₄ pathway of photosynthesis in *Portulaca oleracea* and significance of alanine labelling. Planta 125:273–279.

Hatch, M.D., T. Kagwa and S. Craig, 1975. Subdivision of C₄ pathway species based on differing C₄ acid decarboxylating systems and ultrastructural features. Aust. J. Plant Physiol. 2:111–128.

Hatch, M.D., C.B. Osmond and R.O. Slayter, 1971. Photosynthesis and photorespiration. John Wiley & Sons Inc.

Hess, J.L. and N.E. Tolbert. 1967. Glycolate oxidase in Algae. Plant Physiol. 42:371–379.

Imai, H., S. Iwai and Y. Yamada. 1975. Comparative studies on the photosynthesis of higher plants. IV Further studies on the photosynthetic sugar formation pathway in C₄ plants. Soil Sci. Plant. Nutr. 21:13–19.

Joshi, G., T. Dolan, R. Gee, and P. Saltman. 1962. Sodium chloride effect on dark fixation of CO₂ by marine and Terrestrial plants. Plant Physiol. 37:446–449.

Joshi, G.V., L. Bhosale, B.B. Jamale and B.A. Karadge, 1975. Photosynthetic carbon metabolism in Mangroves. Proc. Int. Symp. Biol and Mgt. of Mangroves, Vol II, 579–594, Hawaii.

Joshi, G.V., 1976. Studies in Photosynthesis under saline conditions. Shivaji University Press, Kolhapur, India.

Joshi, G.V., S.D. Sontakke and L.J. Bhosale, 1980. Studies in photosynthetic enzymes from mangroves. Bot. Mar. 23:745–747.

Joshi, G.V. and A.P. Waghmode, 1980. Photosynthesis in mangroves. *Indian J. Bot.* 4:15–19.

Keller, B. 1925. Halophyten and xerophyten studies. J. Ecology. 13:224 pp.

Kennedy, R.A. 1976. Photorespiration in C₃ and C₄ plant tissue cultures. Significance of Kranz anatomy to low photorespiration in C₄ plants. Plant Physiology 58:573–575.

Kisaki, T. and N.E. Tolbert, 1970. Glycine as a better substrate for photorespiration. Plant Cell Physiol. 11:247–258.

Kluge, M. and C.B. Osmond, 1972. Studies on phosphoenol pyruvate carboxylase and other enzymes of crassulacean acid metabolism of *Bryophyllum tubiflorum* and *Sedum pracaltum*. Z. pflanzenphysiol. 66:97–105.

Ku, S.B. and G.E. Edwards, 1975. Photosynthesis in mesophyll protoplasts and bundle sheath cells of various types of C₄ plants IV. Enzymes of respiratory metabolism and energy utilising enzymes of photosynthetic pathway. Z. *pflanzenphysiol.* 77:16–32.

Kundu, A., P. Palit, R.K. Mandal and S.M. Sircar, 1975. C₃ type photosynthetic carbon fixation in the rice plant. Plant Biochem. J. 3:111–118.

Krause,, G.H., G.H. Lorimer, U. Heber and M.R. Kirk, 1978. Photorespiratory energy dissipation in leaves and chloroplasts. Proc. 4th Int. Cong. on Photosynthesis 1977, 299–310.

Lewis, O.A. and G. Naidoo 1970. Tidal influence on the apparent transpirational rhythms of the white mangrove. S. Afr. J. Sci. 66:268–270.

Lowry, O.H.N.J., Rosenberg, A.L. Farr and R.J. Randall. 1951. Protein measurement with Folinphenol reagent. J. Biol. Chem. 193:265–275.

Lesage, P.M. 1891. Recherches experimentales sur les modifications des feuilles chez les plantes maritimis c.r. Acad. Sci. Paris. 112:672–673.

Long, S.P. and H.W. Woolhouse, 1978. The response of the photosynthesis to light and temperatures in *Spartina townsendii* (sensu lato) a C₄ species from a cool temperate climate J. Expt. Bot. 29:803–814.

Macnae, W. 1968. A general account of the fauna and flora of mangrove swamps and forests in the Indo-West-Pacific region. Adv. Mar. Biol. 6:73–270.

Miller, P.C. 1972. Bioclimate, leaf temperature and primary production in red mangrove canopies in South Florida. Ecol. 53:22–45.

Montfort, C. and U.W. Brandrup, 1927. Physiologische and pflanzengeographische seesalzwirkungen II. Ökologische studien über keimung und erste Entwicklung bei Halophyten. Tp. Wise. Bot. 66–902.

Moore, R.T., P.C. Miller, T. Ehleringer and W. Lawrence, 1973. Season trends in gas exchange characteresitics of three mangrove species. Photosynthetica 7:387–394.

Oliver, D.J. 1979. Mechanism of decarboxylation of Glycine and glycolate by isolated soybean cells. Plant Physiol. 64:1048–1052.

Osmond, C.B. and B. Harris, 1971. Photorespiration during C₄ photosynthesis. Biochim Biophys Acta. 234:270–82.

Pannier, F. 1979. Mangroves implicated by human induced disturbances: A case of study of the Orinoco Delta Mangrove Ecosystem. Environmental Management. 3:205–216.

Raghavendra, A.S. and V.S.R. Das, 1978. Comparative studies on C₄ and C₃ photosynthetic systems. Enzyme levels in the leaves and their distribution in mesophyll and bundle sheath cells. Z. pflanzenphysiol. 87:379–393.

Rehfeld, D.W., D.D. Randall and N.E. Tolbert, 1970.

Enzymes of glycolate pathway in plants without photorespiration. Can J. Bot. 48:1219–26.

Rho, J.H. 1959. Some aspects of the metabolism of a marine diatom *Nitzehia closterium*. International Oceanographic Conference. Reprints Washington. 199 pp.

Schimper, A.F.W. 1891. Die indo-malaxische strandflora. Bot. Mitt. aus den Trop. Vol. 3 Jena. 204 pp.

Shomer Ilan, A. and Y. Waisel, 1973. The effect of sodium chloride on the balance between the C_3 and C_4 carbon fixation pathways. Physiol Plant. 29:190–193.

Shomer Ilan, A.S. Beer and Y. Waisel, 1975. *Suaeda monoica* a C_4 plant without typical bundle sheaths. Plant Physiol. 56:676–679.

Shomer Ilan, A., Neumann Ganmore and Y. Waisel, 1979. Biochemical specialization of photosynthetic cell layers and carbon flow paths in *Suaedo monoica*. Plant Physiol. 64:963–965.

Stoddard, E.M. 1965. Identifying plants by epidermal characters. Can Agric. Exp. Sta. Circular 227.

Teas, H.J. 1976. Productivity of Biscayne Bay mangroves. Biscayne Bay Symposium I. Univ. Miami. Special report No. 5, 103–111.

Teas, H.J. 1979. Silviculture with saline water. The Biosaline concept. Edited by Alexander Hollaender (Plenum publ. corporation). pp. 117–161.

UNESCO, Panel Report (Mangroves) 1976. Meeting held at Phuket Thailand.

Walsh, G.E. 1974. Mangroves: a review in Ecology of Halophytes edited by Reimold and Queen. Academic Press. 51 pp.

Waghmode, A.P. and G.V. Joshi, 1979. Glycolate oxidase in halophytes. Indian J. Expt. Biol. 17:111–112.

Warik, R.P. 1960. Physiological and ecological studies of halophytes. Ph. D. Thesis, University of Bombay (India).

Webb, K.L. and J.W.A. Burley. 1965. Dark fixation of CO_2 by obligate and facultative salt marsh halophytes. Can. J. Bot. 43:281–285.

Woo, K.C. and C.B. Osmond, 1976, Glycine decarboxylation in mitochandria isolated from Spinch leaves. Aust. J. Plant Physiol, 3:771–785.

Zelitch, I. 1972. Comparison of the effectiveness of glycolics acid and glycine as substrate for photorespiration. Plant Physiol. 50:109–113.

CHAPTER 2

Photosynthetic gas exchange properties and carbon isotope ratios of some mangroves in North Queensland

T.J. ANDREWS, B.F. CLOUGH and G.J. MULLER

Australian Institute of Marine Science, P.M.B. No. 3, Townsville, M.S.O. Q. 4810. Australia

Abstract. The gas exchange properties of fully expanded leaves on mature trees of *Rhizophora stylosa* and *R. apiculata* at Hinchinbrook Island, North Queensland were studied using a measurement system based on infrared gas analysers and dew point hygrometers. In some cases, CO_2 and water vapour exchange were followed in diurnal 'tracking' experiments under natural illumination. Leaf temperature and the vapour pressure deficit inside the leaf chamber were maintained under changing conditions which simulated those experienced by other leaves in the canopy at the time. These experiments revealed that leaf temperature was a key factor controlling assimilation under natural conditions.

In other experiments, the response of gas exchange parameters to variation in CO_2 partial pressure, light flux and leaf temperature under otherwise constant conditions were studied using artificial illumination. The stomata were found to be very sensitive to changes in environmental conditions, irrespective of whether these changed when simulating canopy conditions or in response to experimental manipulation. In both cases there appeared to be parallel changes in stomatal conductance and assimilation rate so that the calculated CO_2 partial pressure in the intercellular spaces was maintained more or less constant. The only case when this nexus between conductance and assimilation rate was not complete occurred when CO_2 partial pressures were manipulated. Maximum rates of assimilation in normal air were about 8 micromol $m^{-2} s^{-1}$ (12.6 mg $dm^{-2} hr^{-1}$) and occurred at a leaf temperature of about 30°C and light quantum fluxes above 700 micromol $m^{-2} s^{-1}$.

The photosynthetic response of *Rhizophora* leaves to light, CO_2 partial pressure and temperature were typical of the C_3 photosynthetic pathway. Additional evidence for the operation of the C_3 pathway in mangroves was obtained from an analysis of the isotopic discrimination against ^{13}C by 16 species from 14 genera of North Queensland mangroves. The degree of discrimination against ^{13}C, as represented by the $\delta^{13}C$ value, ranged from $-32.2‰$ to $-24.6‰$. These values are typical of C_3 plants and clearly more negative than the range for C_4 plants. These data lead to the conclusion that, in common with most trees, mangroves photosynthesize via the C_3 pathway.

Introduction

Mangroves attain their greatest diversity and luxuriance in the humid tropics where solar radiation and air temperatures are commonly high. These factors combine to ensure that leaves exposed to direct sunlight often experience leaf temperatures which are supra-optimal for C_3 photosynthesis. Under these conditions adaptations which lead to a reduction in leaf temperature or which modify the photosynthetic characteristics of leaves would clearly be an advantage.

H.J. Teas (ed), Physiology and management of mangroves.
© *1984 Dr W. Junk Publishers, The Hague. ISBN 90 6193 949 6. Printed in the Netherlands.*

Previous studies of the gas exchange characteristics of mangrove leaves (Moore *et al.*, 1972; 1973) support the view that the C_3 pathway is the primary photosynthetic pathway in mangroves. The results of Joshi and co-workers (1974; 1976), however, seem to be at variance with this view.

Recently, a number of studies on mangroves in northern Australia have been initiated with the aim of developing an understanding of the basic physiology which underlies their success in exploiting the intertidal zone. In this paper we consider some aspects of the gas exchange characteristics of mangrove leaves, measured both in the laboratory and in the field.

Methods

Field studies were carried out at Hinchinbrook Island, North Queensland (18° 15' S). The site was located in a mixed stand of *Rhizophora apiculata, R. stylosa* and *Bruguiera gymnorhiza* along the fringe of a large tidal creek.

Measurements of gas exchange were made on fully expanded leaves of *R. apiculata* and *R. stylosa* at the top of the canopy about 10 m above the ground. Gas exchange by a single leaf enclosed in a ell-ventilated chamber, constructed of aluminium and glass, was measured with infrared gas analysers (CO_2), dew point hydrometers (water vapor) and a mass flow meter (air flow). For some experiments the gas exchange system was used in an open-ended mode and for others in a closed configuration. Leaf temperature was measured with two fine-gauge copper-constantin thermocouples fixed with transparent tape to the upper surface of the leaf close to either end. The photon flux density of photosynthetically active radiation (400–700 nm) inside the chamber was measured with a Lambda quantum sensor. Leaf temperature and the partial pressures of CO_2 and water vapor inside the chamber were controlled using the appropriate transducers and a small computer. Under steady state conditions it was possible to control leaf temperature to $\pm 0.2°$ C, pH_2O to ± 0.2 mbar and pCO_2 to $\pm 2 \mu$bar (approx. ± 2 ppm). Under non-steady state conditions (i.e. when forcing conditions inside the chamber to track those prevailing outside)

the degree of control was in some cases slightly worse (see Fig. 1). The computer was also used for data acquisition, providing an on-the-spot output of both primary and derived data. This allowed experimental procedures and conditions to be varied immediately.

In some experiments, CO_2 and water vapor exchange were followed in 'diurnal tracking experiments' under natural illumination, in which leaf temperature and water vapor pressure inside the chamber were controlled to simulate conditions measured with an independent set of sensors placed on or near other leaves in the canopy outside the leaf chamber. In other experiments, the response of photosynthesis and transpiration to variation in the CO_2 parital pressure, light flux and leaf temperature, under otherwise constant conditions, were studied using artificial illumination provided by a multi-metal halide discharge lamp. PAR fluxes in the range 0–1700 μmol m^{-2} s^{-1} were obtained by interposing filters (Sarlon shade cloth) between the light source and the leaf chamber.

In addition to the field studies, an exploratory study of the gas exchange characteristics of leaves on 18-month old seedlings of *R. apiculata* was carried out in the laboratory. The seedlings for this study were grown in a shadehouse covered with 50% shade cloth.

For the determination of $\delta^{13}C$ values, samples of mangrove leaves from the Hinchinbrook Island area were collected, dried at 80° C, and ground to a powder. $\delta^{13}C$ values of this material were determined in the laboratory of Dr. C.B. Osmond at the Australian National University using established methods (Smith and Epstein, 1971).

Results

Tracking experiments

Tracking experiments have been carried out on leaves of *R. stylosa* over a wide range of environmental contitions. These include cloudy and clear days, and the use of reference leaves outside the chamber at their natural inclination or artificially held horizontal. Some experiments under natural

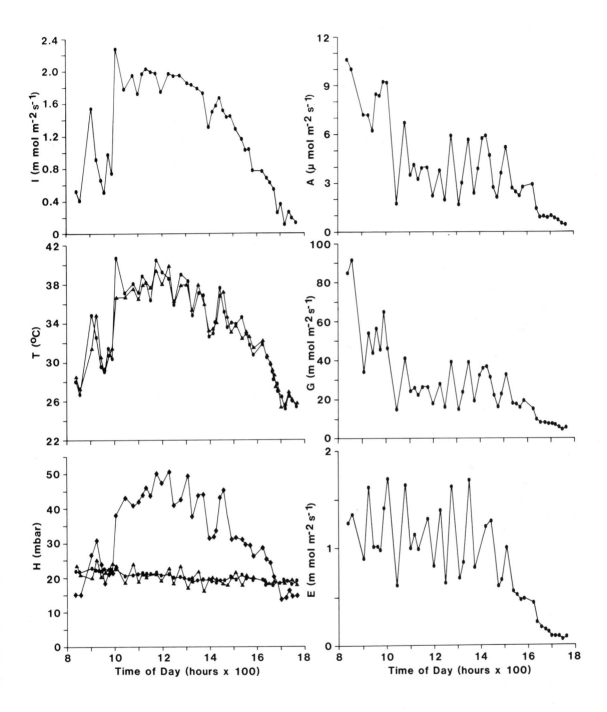

Fig. 1. Changes in irradiance, vapour pressure, leaf temperature, and measured gas exchange parameters in an exposed 'sun' leaf of *Rhizophora stylosa* in the field during the course of a day. I, photon flux density (400–700 nm); T, leaf temperature (●, reference leaf outside chamber; ▲, leaf inside chamber); H, vapour pressure (●, ambient outside chamber; ▲, inside chamber; ◆, vapour pressure difference between leaf and air inside chamber); A, assimilation rate; G, leaf conductance; E, evaporation rate.

illumination were also performed where the leaf temperature and water vapour pressure were maintained constant.

The data obtained from these experiments were often difficult to interpret because of intermittent cloud and the consequent marked fluctuation in isolation. An example of the kind of results obtained are shown in Fig. 1, where data for one of the less cloudy days are presented. These data were obtained using a reference leaf artificially held horizontal outside the chamber and a CO_2 partial pressure of $400\,\mu$bars. The data are particularly useful in revealing the interactive relationships between irradiance, leaf temperature, stomatal conductance, net photosynthesis and evaporation.

At low light levels, and correspondingly low leaf temperatures, early in the morning, stomatal conductance and net photosynthesis were both high. As the day progressed, however, the increasing irradiance led to a marked increase in leaf temperature coupled with parallel reductions in stomatal conductance and net photosynthesis. In the latter

part of the day irradiance decreased, with a corresponding reduction in leaf temperature. However, there was not a concomitant increase in stomatal conductance or net photosynthesis. A reduction in stomatal conductance and net photosynthesis late in the afternoon has been observed both in 'tracking' experiments and in experiments where the leaf in the chamber had been subjected to constant conditions. The reason for this effect has yet to be investigated.

These results illustrate several important features of the gas exchange characteristics of mangroves:

1 As expected, there is a close positive correlation between leaf temperature and irradiance.
2 Stomatal conductance is closely and inversely related to leaf temperature.
3 Net photosynthesis is markedly influenced by leaf temperature. Changes in net photosynthesis in response to rapid fluctuations in leaf temperature are greater than can be explained by changes in

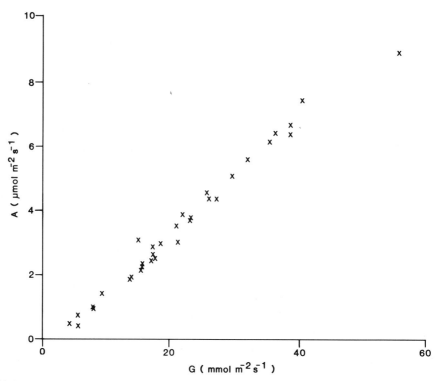

Fig. 2. Relationship between assimilation rate and stomatal conductance in an exposed 'sun' leaf of *Rhizophora stylosa* under 'tracking' conditions in the field.

18

stomatal conductance, however, and a direct effect of leaf temperature on net photosynthesis is indicated.

Naturally, the vapor pressure difference between the leaf and air is also influenced to a major degree by leaf temperature and, in this instance, vapor pressure deficits of as high as 50μbar were recorded.

An interesting feature of the data is that they show that the inverse relationship between leaf temperature and stomatal conductance results in a relatively constant rate of evaporation throughout most of the day (Fig. 1).

When the temperature of the leaf in the chamber was forced to track the diurnal temperature fluctuations experienced by a leaf allowed to adopt its natural, nearly vertical inclination, leaf temperatures were typically considerably lower than those shown in Fig. 1 and were higher at mid-morning and mid-afternoon than at mid-day. Water vapor deficits also were concomitantly lower. Under these conditions assimilation and stomatal conductance remained at high levels throughout most of the day provided illumination was adequate (data not shown).

A notable feature of the data for all tracking experiments is the strong positive correlation between assimilation and stomatal conductance. Fig. 2 shows this correlation for the same day as Fig. 1.

Gas exchange characteristics of Rhizophora apiculata

It was evident from the 'tracking' experiments that the stomata were remarkably responsive to small increases in leaf temperature when the latter was in the supra-optimal range. This was also found in the course of the experiments described below. By contrast, the response of the stomata to a stepwise reduction or increase in light was slower, and periods of up to 30 minutes at a given photon flux density were often required before the stomatal conductance settled to a relatively constant value. Because of the time required to obtain a light response curve (usually 1 day) at a constant external CO_2 partial pressure and leaf temperature, only a

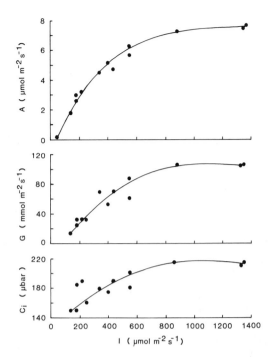

Fig. 3. Effect of light (I, 400–700 nm) on assimilation rate (A), stomatal conductance (G) and intercellular CO_2 partial pressure (C_i) in a 'sun' leaf of *Rhizophora apiculata*.

few full light response curves were attempted. The more usual procedure was to determine a CO_2 response curve for each of 3 or 4 light levels at a constant temperature over the course of a day.

A typical light response curve for *R. apiculata* is shown in Fig. 3. At low light levels net photosynthesis increased more or less linearly with photon flux density, giving an incident quantum efficiency of about 0.017 and a light compensation point of about 30μmol photons $m^{-2} s^{-1}$. There was little increase in net photosynthesis beyond a photon flux density of 700μmol $m^{-2} s^{-1}$. These characteristics varied slightly from leaf to leaf, presumably as a result of differences in age, orientation and degree of exposure of the leaf, but these aspects have yet to be investigated in detail. The characteristics of the light response curve shown in Fig. 3 are quite unexceptional, and are similar to those in the literature for mangroves (Moore *et al.*, 1972; 1973; Attiwill and Clough, 1980) and for other trees (e.g. Hesketh and Baker, 1967; Larcher, 1969).

A feature of the data shown in Fig. 3 is that the

decline in assimilation at low light fluxes is not due to stomatal closure. Whereas the stomatal conductance fell from about $105\,\mu$mol m^{-2}s^{-1} (resistance = 3.8 s cm^{-1}) to around $15\,\mu$mol m^{-2}s^{-1} (resistance = 26.7 s cm^{-1}) as the quantum flux density was reduced from 1350 to $180\,\mu$mol m^{-2}s^{-1}, the reduction was roughly proportional to the reduction in assimilation so that the CO_2 partial pressure inside the leaf (C_i) was reduced only slightly. This is consistent with the correlation between assimilation and stomatal conductance mentioned earlier (Fig. 2) and provides evidence for the existence of a feedback loop between the rate of assimilation and stomatal aperture which, at a normal atmospheric partial pressure of about $330\,\mu$bars, operates to maintain a more or less constant CO_2 partial pressure inside the leaf (in this case $150–220\,\mu$bars) (Wong et al., 1979). Further evidence for this was obtained from laboratory-based measurements on leaves of 18-month old seedlings. Some results from these experiments are presented in Table 1. These show that whereas C_i varied little with irradiance, the efficiency of carboxylation or mesophyll conductance (given by the relationship $A/[C_i - \Gamma]$) was highly light dependent.

A typical CO_2 response curve for a 'sun' leaf of R. apiculata in the field at a leaf temperature of 30°C and at a high quantum flux density of $1400\,\mu$mol m^{-2} s^{-1} is shown in Fig. 4. Since the stomatal conductance to some extent varies inversely with the external CO_2 partial pressure, net

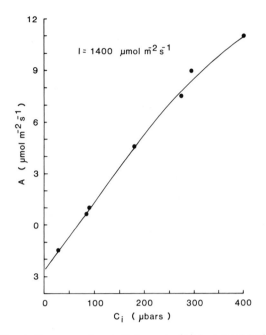

Fig. 4. Response of assimilation rate (A) in a 'sun' leaf of Rhizophora apiculata as a function of the intercellular CO_2 partial pressure (C_i) at a high light level ($I = 1400\,\mu$mol photons m^{-2}s^{-1}, 400–700 nm).

photosynthesis has been plotted against the CO_2 partial pressure inside the leaf (C_i). As expected, net photosynthesis initially increased almost linearly with C_i, the relationship apparently becoming progressively more non-linear at values of C_i beyond $200\,\mu$bars. The CO_2 compensation point (Γ) derived from the curve was about $60\,\mu$bars, and

Table 1. Effects of CO_2 partial pressure and irradiance on some photosynthetic characteristics of 18-month old Rhizophora apiculata seedlings. C_a, ambient CO_2 partial pressure; I, quantum flux density (400–700 nm); A, assimilation rate; G, stomatal conductance; C_i, intercellular CO_2 partial pressure; $A/(C_i - \Gamma)$, carboxylation efficiency. Measurements were made at a leaf temperature of 28°C.

Ca (μbars)	I (μmol m^{-2}s^{-1})	A (μmol m^{-2}s^{-1})	G (mmol m^{-2}s^{-1})	C_i (μbars)	$A/(C_i - \Gamma)$*
660	1350	12.7	90	425	34.8
	600	12.5	78	400	36.8
330	1350	6.2	61	160	62.0
	600	5.8	60	175	50.0
	265	3.8	43	190	29.2
165	1400	2.8	80	105	62.2

* nmol m^{-2}s^{-1} μbar^{-1}

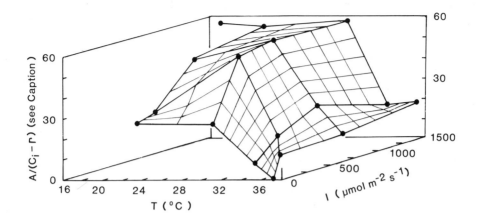

Fig. 5. Three-dimensional representation of the effect of leaf temperature (T) and light (I, 400–700 nm) on the carboxylation efficiency ($A/[C_i-\Gamma]$) in a 'sun' leaf of *Rhizophora apiculata*. $\Gamma = CO_2$ compensation point. Units for the carboxylation efficiency are nmol $m^{-2} s^{-1}$ μbar^{-1}.

the carboxylation efficiency, as defined by the initial slope of the relationship between assimilation and C_i (i.e. $A/[C_i - \Gamma]$), was 41 nmol $m^{-2} s^{-1}$ μbar^{-1}. These characteristics are entirely consistent with the C_3 photosynthetic pathway.

The effects of leaf temperature and irradiance on the photosynthetic capacity of exposed 'sun' leaves of *R. apiculata* in the field are summarised in Fig. 5 and Table 2. Each datum point in Fig. 5 was derived from the initial slope of CO_2 response curves at the specified leaf temperature and light level. All data points, and particularly those at high leaf temperature, where both assimilation and stomatal conductance were very low, are subject to error, but the overall trends in photosynthetic capacity in response to leaf temperature and light are nevertheless quite clear. These are: *1.* The carboxylation efficiency ($A/[C_i - \Gamma]$) increased slightly with leaf temperature from 17 to 30° C, but thereafter dropped sharply, falling to close to zero at 37° C (Fig. 5). *2.* The CO_2 compensation point (Γ) tended to increase with leaf temperature (Table 2). *3.* The carboxylation efficiency tended to rise with increasing light (Fig. 5). *4.* Stomatal conductance tends to decrease with increasing temperature above 30° C and with decreasing light levels (Table 3). The stomatal conductances at 30° C are for some reason anomalous; many other measurements at this temperature indicate a maximum

stomatal conductance at high light levels of around 105 nmol $m^{-2} s^{-1}$. These trends suggest that a leaf temperature of 22–30° C is optimal for assimilation in *R. apiculata*. Results from the tracking experiments described earlier indicate a similar leaf temperature optimum for assimilation in 'sun' leaves of *R. stylosa*.

$\delta^{13}C$ values for mangrove leaves

The $\delta^{13}C$ values for mangroves ranged from -25.7% to -32.2% (Table 4). These values are typical of C_3 plants and clearly more negative than the range for C_4 plants (Smith and Epstein, 1971). Since it reflects discrimination against ^{13}C by the initial carboyxlation reaction of photosynthesis, the $\delta^{13}C$ value has proved to be a most reliable

Table 2. Effect of irradiance (400–700 nm) and leaf temperature on the CO_2 compensation point ($\mu bars$) of exposed 'sun' leaves of *Rhizophora apiculata*.

Quantum flux Density ($\mu mol\ m^{-2}s^{-1}$)	Leaf Temperature (°C)				
	17	22	30	34	37
175	–	150(?)	55	65	–
400	–	80	70	55	95
750	–	60	65	65	80
1500	40	50	65	70	90

Table 3. Effects of leaf temperature and irradiance (400–700 nm) on somatal conductance (mmol m^{-2}s^{-1}) in exposed 'sun' leaves of *Rhizophora apiculata*.

Photon Flux Density (μmol m^{-2}s^{-1})	Leaf Temperature (°C)				
	17	22	30	34	37
175	–	27	48	19	9
400	–	65	40	35	16
750	–	55	43	35	22
1500	110	110	57	36	34

Table 4. δ^{13}C values for a number of species of mangroves.

Species	δ^{13}C Value (‰)
Acanthus ilificolius L.	– 27.6
Aegialitis annulata R. Br.	– 28.7
Aegiceras corniculatum (L.) Blanco	– 28.4
Avicennia eucalyptifolia Zip. ex. Mig.	– 26.9
Bruguiera gymnorhiza (L.) Lam.	– 27.7
Bruguiera parviflora (Roxb.) Wight & Arn.	– 31.9
Ceriops tagal var *tagal* (Perr.) C.B. Rob.	– 32.2
Excoecaria agallocha L.	– 27.3
Heritiera littoralis Ait.	– 29.4
Lumnitzera littorea (Jack.) Voigt	– 28.7
Lumnitzera racemosa Willd.	– 26.0
Osbornea octodonta F. Muell.	– 26.4
Rhizophora apiculata Bl.	– 9.2
Scyphiphora hydrophyllacea Gaertn.	– 26.8
Sonneratia alba J. Sm.	– 26.7
Xylocarpus australisicum Ridl.	– 26.3
Range for C$_4$ plants	– 11.5 to – 18.3*
Range for C$_3$ plants	– 23.2 to – 34.3*

* Smith and Epstein, 1971.

character for distinguishing between C$_3$ and C$_4$ photosynthesis in terrestrial higher plants.

Discussion

The results presented above show that high leaf temperatures have a profound effect on both stomatal conductance and the efficiency of carboxylation. The two effects acting in concert cause a large reduction in photosynthetic carbon assimilation at leaf temperatures only a few degrees above what appears to be an optimum leaf temperature

for photosynthesis of 30° C. The temperature of 'sun' leaves experimentally held normal to the incoming solar radiation commonly exceeds 35° C and may reach 45° C in North Queensland. By contrast, exposed leaves which are steeply inclined have temperatures only a few degrees higher than air temperature. This emphasises the critical importance of the nearly vertical orientation of 'sun' leaves in the Rhizophoraceae as an adaptation for minimising the incoming radiation load and hence leaf temperature. The physiological cost, in terms of photosynthetic carbon assimilation, is high indeed for any leaf unlucky enough to find itself normal to the sun's rays.

While the nearly vertical orientation of the 'sun' leaves of the Rhizophoraceae clearly is crucial to the maintenance of an acceptable heat balance, there is some loss of potential photosynthetic production because of the lower quantum flux densities received by the leaf. Our results indicate, however, that the reduction in assimilation as a consequence of low light levels is much less than that due to high leaf temperatures. Both the light response curve (Fig. 3) and the data in Table 1 show that the rate of assimilation falls by less than 20% for a decrease in the quantum flux density of 50% from 1350 to 700 μmol m^{-2} s^{-1}. Clearly the optimum leaf angle required to maximise photosynthetic production must represent a compromise between the need to avoid excessively high leaf temperatures on one hand and, on the other, to absorb enough quanta to maintain as high a photosynthetic rate as possible. Such considerations are not restricted to the Rhizophoraceae, or to mangroves growing in warmer climates. A similar response to high radiation loads and photosynthetically active radiation has been found in leaves of *Avicennia marina* growing in Westernport Bay, where it is near its southernmost limit (Attiwill and Clough, 1980).

The gas exchange characteristics of *Rhizophora* leaves observed in this study are in substantial agreement with those of Moore *et al.* (1972; 1973) for *Rhizophora, Avicennia* and *Laguncularia,* and with those of Kimball and Cowan (1979), and Attiwill and Clough (1980) for *Avicennia.* Photosynthesis typically is saturated at photon flux densities of one-half to two-thirds full sunlight and has a temperature optimum below 35°C. A CO_2 compensation point of about $60\,\mu$bars at the optimum temperature for leaf photosynthesis is indicative of photorespiratory activity. These features, together with the $\delta^{13}C$ values ranging form -25.7% to -32.2%, are clearly C_3 characters. We therefore conclude that mangroves, in common with most tree species, employ C_3 photosynthetic biochemistry.

Literature cited

Attiwill, P.M. and B.F. Clough. 1980. Studies of gas exchange in the white mangrove. Photosynthetica 14:40–47.

Hesketh, J. and D. Baker. 1967. Light and carbon assimilation by plant communities. Crop Sci. 7:285–293.

Joshi, G.V. 1976. Studies in Photosynthesis under Saline Conditions. Shivaji University Press, Kolhapur, India.

Joshi, G.V., M.D. Karekar, C.A. Gowda and L. Bhosale. 1974. Photosynthetic carbon metabolism and carboxylating enzymes in algae and mangrove under saline conditions. Photosynthetica 8:51–52.

Kimball, M.C. and I.R. Cowan. 1979. Assimilation rate and leaf conductance in seedlings of *Avicennia marina*. Paper presented at the annual meeting of the Australian Society of Plant Physiologists, Canberra, 1979.

Larcher, W. 1969. The effect of environmental and physiological variables on the carbon dioxide gas exchange of trees. Photosynthetica 7:167–198.

Moore, R.T., P.C. Miller, D. Albright and L.L. Tieszen. 1972. Comparative gas exchange characteristics of three mangrove species in winter. Photosynthetica 6:387–393.

Moore, R.T., P.C. Miller, J. Ehleringer and W. Lawrence, 1973. Seasonal trends in gas exchange characteristics of three mangrove species. Photosynthetica 7:387–394.

Smith, B.N. and S. Epstein. 1971. Two categories of $^{13}C/^{12}C$ ratios for higher plants. Plant Physiol. 47:380–384.

Wong, S.C., I.R. Cowan and G.D. Farquhar. 1979. Stomatal conductance correlates with photosynthetic capacity. Nature 282:425–426.

CHAPTER 3

Productivity and phenology of *Avicennia marina* (Forsk.) Vierh. and *Bruguiera gymnorrhiza* (L.) Lam. in Mgeni estuary, South Africa

T.D. STEINKE and L.M. CHARLES

Department of Botany, University of Durban-Westville, Durban, South Africa

Abstract. Estimates of productivity of mangroves were obtained with the use of litter baskets placed at random in two communities: one dominated by *Avicennia marina* (Forsk.) Vierh. and the other comprising an almost pure stand of *Bruguiera gymnorrhiza* (L.) Lam. Average litter production for the *Avicennia* community was 2.61 g and 1.31 g dry matter m^{-2} day^{-1}, while that for the *Bruguiera* stand was 2.59 g and 2.14 g dry matter m^{-2} day^{-1}, for the 1978/79 and 1979/80 harvests respectively. Highest values in both communities were recorded at the time of seedling abscission in autumn. Woody material formed a relatively low proportion of the total litter yields. In the *Avicennia* community leaf fall was slightly higher in the dry, cool months. In the *Bruguiera* stand leaf appearance was high and leaf fall low in the wet, warmer months, while during the dry, cool period figures for leaf fall were higher than those for leaf appearance. The significance of the results is discussed in terms of the distribution of the communities in a sub-tropical region approaching the southernmost limit of these mangroves.

Introduction

In southern Africa mangroves extend as far south as East London (33° S), although they are clearly of negligible importance as an estuarine community at this southern limit of their range. Further northwards mangrove communities may be relatively well-developed and are considered to play a significant role in the ecology of the many estuaries along which they occur (Moll, Ward, Steinke & Cooper, 1971). However, there is no reliable information available on the productivity of these mangroves, and in this respect it is therefore difficult to ascribe more than a tentative role to them at this stage. On the other hand, tropical and subtropical communities in America (Lugo & Snedaker, 1974; Pool, Lugo & Snedaker, 1974), in Australia (Bunt, 1979) and in Thailand (Christensen, 1978) have in recent years been the subject of extensive research, and

the extent of the contribution of these communities to an estuarine system consequently can be determined with greater reliability. For this reason a project to assess the productivity of the two most important mangrove species in southern Africa, *Avicennia marina* (Forsk.) Vierh. and *Bruguiera gymnorrhiza* (L.) Lam., was initiated at Beachwood, Mgeni estuary, by recording litter fall. While recording litter fall is valuable as an indicator of mangrove productivity, it is appreciated that it does not reflect the full extent of net photosynthetic yield as other important components of production (viz. major stems, roots) are neglected. This project had the additional objective of supplementing the limited phenological data presently available on these southern mangroves. The results have shown that, while mangroves do make a significant contribution towards the detrital food chain in the Beachwood area, the productivity of

H.J. Teas (ed), Physiology and management of mangroves.
© *1984 Dr W. Junk Publishers, The Hague. ISBN 90 6193 949 6. Printed in the Netherlands.*

this community is low in relation to tropical communities. Although the project has been running for more than two years, in this paper only the results obtained in the first two years (1978/79 and 1979/80) will be considered.

Methods

The study was conducted in the Beachwood mangrove swamp at the smouth of the Mgeni River which enters the sea at Durban. The mangrove and other plant communities in this area have recently been described by Padia (1980). The Beachwood stream, which joins the Mgeni River near its mouth, lies roughly in a NE-SW direction, parallel to the coast, and is sandwiched between a range of sand dunes and the urban development to the landward side. The Beachwood area, which is long and narrow (approximately 2,60 × 0,35 km), has been declared a nature reserve. Mangroves occupy only 29 hectares of the total area. The main mangrove species are *Bruguiera gymnorrhiza* and *Avicennia marina*, although *Rhizophora mucronata* Lam. is also present. The mangrove swamp is flooded entirely only by spring tides and there is a transition to

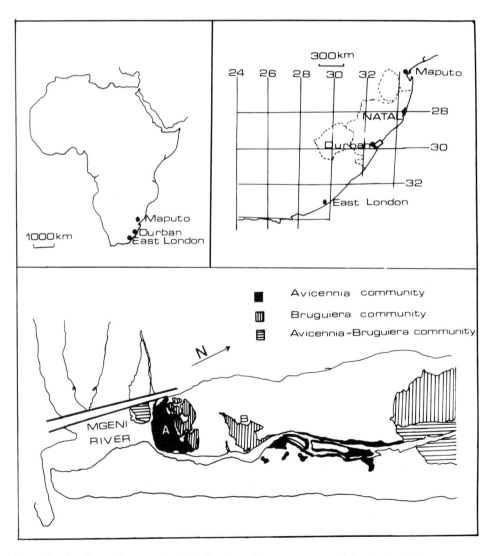

Fig. 1. Mgeni River, showing the southern section of the Beachwood mangrove swamp (after Padia, 1980).

fresh water conditions in the northern part.

Two areas near the mouth were chosen for this study: a stand dominated by *A. marina* (A) and an almost pure stand of *B. gymnorrhiza* (B) (Fig. 1) These will for convenience be referred to as the *Avicennia* and *Bruguiera* communities respectively. The *Avicennia* community (1,25 hectare in extent) comprised scattered tall trees (9 m) with increasing numbers of *Bruguiera* (7 m) towards the west and north where the latter form a dense fringe. An occasional tree or sapling of *R. mucronata* is also present. The *Bruguiera* community (0,67 hectare) showed a decrease in height from

7 m near the water's edge to 5 m at the landward margin.

In the *Avicennia* and *Bruguiera* communities 15 and 10 litter baskets respectively, each measuring 0,25 m² in area, were placed at random. Each area was stratified and a basket positioned by choice of random numbers within each stratum. The baskets were placed in position towards the end of 1978 and the project is still in progress. Collections of litter were made fortnightly. The harvested material was separated into the various plant components of each species, oven-dried at 70° C for a week, and then weighed. It was not possible to separate *Avi-*

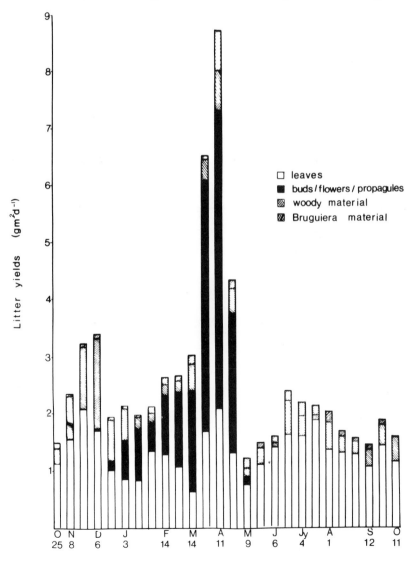

Fig. 2. Litter fall in *Avicennia* community in 1978/79 harvest year.

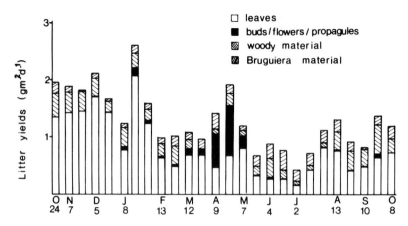

Fig. 3. Litter fall in *Avicennia* community in 1979/80 harvest year.

cennia buds, flowers and propagules which are therefore considered together. The abscised leaves and stipules of *Bruguiera* from each harvest were counted and the number of stipules used as an indication of leaf appearance after field observations had confirmed the validity of this assumption. Root samples were not collected.

The climate of Durban is subtropical. The mean annual temperature is 21°C, February having a mean temperatre of 25.2°C and July a mean temperature of 16.7°C are the hottest and coldest

months of he year respectively. Mean relative humidity for the year is 73 per cent and mean annual rainfail is 1013 mm, of which 66 per cent is received in the months October to March (S.A. Weather Bureau).

Results and discussion

Avicennia community

During 1979/80 total litter fall for this community

Tabel 1. Mean dry matter yields of litter components (g m^{-2} d^{-1}) in two mangrove communities at Beachwood.

Communities	Harvest years	Species	Leaves	Stems	Stipules	Calyces	Buds	Flowers	Propagules	Total
Bruguiera gymnorrhiza	1978/79	Bruguiera	1.66	0.10	0.13	0.14	–	0.03	0.50	2.59
		Avicennia	0.02	0	–	–		0.01		
	1979/80	Bruguiera	1.26	0.05	0.11	0.11	–	0.03	0.57	2.14
		Avicennia	0.01	0	–	–		0		
Avicennia marina	1978/79	Bruguiera	0.08	·0	0.01	0	–	0	0.02	2.61
		Avicennia	1.34	0.43	–	–		0.73		
	1979/80	Bruguiera	0.12	0	0.01	0.01	–	0	0.02	1.31
		Avicennia	0.82	0.24	–	–		0.09		

remained fairly constant for most of the year, although there was a marked increase from mid-March to the end of April largely as a result of the fall of propagules (Fig. 2). In the second year of the experiment litter fall showed a decline in latter months with only a low peak in April as a result of greatly reduced propagule production (Fig. 3). Relatively high yields of litter were obtained on 22 November 1978 and 6 December 1978, apparently due to unseasonally strong winds which followed a period of wet, warm weather. The importance of wet, windy weather in determining litter fall has also been observed by Pool, Lugo & Snedaker (1974). Those high winds, which caused branches to be broken off the tree, are probably also responsible for the fact that stem litter comprised a relatively high proportion of the total yield at this time. The contribution of stem material fluctuated but was seldom high. In this community stems made up not more than 16.5% of the total litter and for *A. marina* alone the litter: wood ratio was approximately 4.8 : 1.0 (Table 1). Much of the stem material consisted of broken twigs, small branches, and flakes of bark. These flakes of corky material on the trunk and larger branches of *A. marina* are characteristic and appeared to contribute an almost constant amount of material to the stem litter yield.

The contribution of *B. gymnorrhiza* to the litter is significant at the peak of the period of propagule abscission, although throughout the year small amounts of leaves and stipules were also collected in the litter baskets (Figs. 2 and 3). In the 1979/80 harvest period *Avicennia* litter yields were so low that the contribution by *Bruguiera* formed a significant proportion of the total litter recorded (Fig. 3).

During 1978/79 leaf fall was fairly constant, although slightly higher values were recorded in the dry, cool months when the average collection was 1.51 g m^{-2} d^{-1} as compared with 1.16 g m^{-2} d^{-1} in summer (Fig. 4). The high leaf drop in November/early December 1978 has been attributed to prolonged wet conditions followed by strong winds. In 1979/80 the position was reversed in that the average collections were 1.12 and 0.53 g m^{-2} d^{-1} for the warm and cool, dry months respectively (Fig. 5). Although research is in progress to determine the longevity of leaves, no information is presently

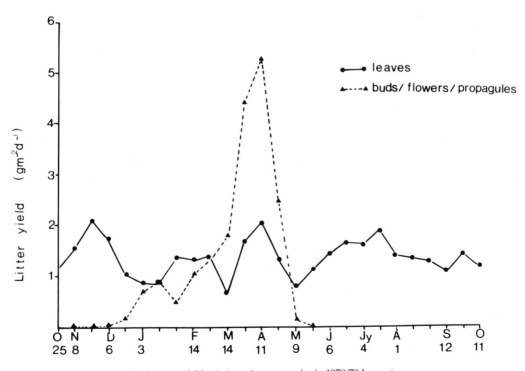

Fig. 4. Litter yields of leaves and reproductive material in *Avicennia* community in 1978/79 harvest year.

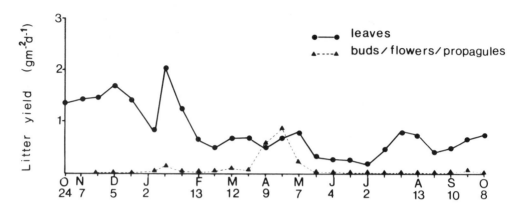

Fig. 5. Litter yields of leaves and reproductive material in *Avicennia* community in 1979/80 harvest year.

available, but there are indications from work on *A. germinans* that leaves may be present on the trees for approximately one year before abscission (Teas, 1976).

In the 1978/79 harvests bud fall was recorded as early as 8 November, and this coincided with the appearance of buds on the trees, so bud fall can be taken as a reliable indication of time of bud formation. This applies similarly to flowering. Flowers made their first appearance in litter baskets on 6 December, although buds formed the greater part of the yield at that stage. However, by 20 Decem-

ber flowers were more common in the litter. Flowers were present until the beginning of May, but the fact that propagules are usually not found on the trees later than this, suggests that flowers formed so late, develop no further and abscise. On 17 January the first small propagules were collected in the baskets. Although they were small, indications from other work (Steinke, unpublished data) suggest that under favourable conditions propagules of that size could survive and establish sucessfully. From this date propagules formed an increasingly important fraction of the total litter

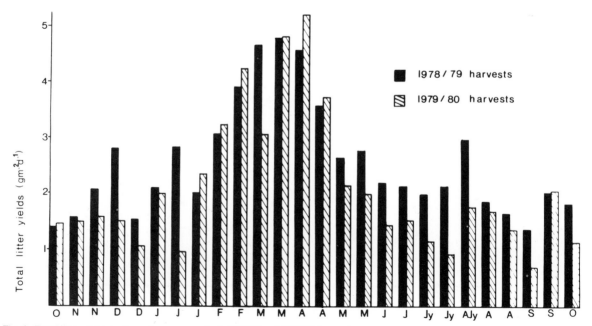

Fig. 6. Total litter yields in *Bruguiera* community in 1978/79 and 1978/80 harvest years.

yield and by early March formed almost the entire yield of reproductive material. On 28 March and 11 April propagules comprised 68 per cent and 65 per cent respectively of the total litter yield. After the slow initial increase in yield of reproductive parts there was a steep increase as propagules formed and fell, leading to a maximum on 11 April. Thereafter there was a sharp abatement with decreasing yields of reproductive material recorded until the end of May (Fig. 4). In the following year a similar pattern was revealed, although propagule production at its peak on 23 April comprised only 45 per

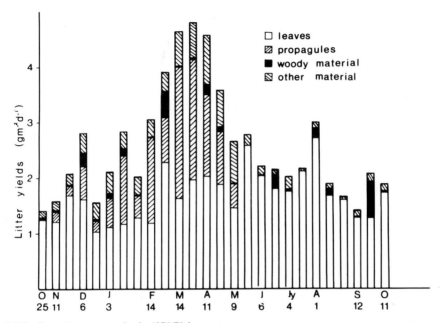

Fig. 7. Litter fall in *Bruguiera* community in 1978/79 harvest year.

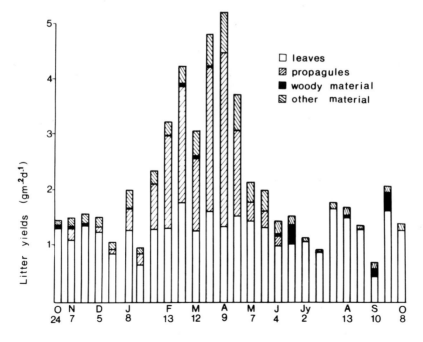

Fig. 8. Litter fall in *Bruguiera* community in 1979/80 harvest year.

31

cent of the total litter yield (Fig. 5). It is perhaps significant that leaf litter yields were also lower, but more information is necessary before a definite relationship can be established. Investigations into possible hormonal involvement in controlling both propagule and leaf abscission have been initiated.

Of the total litter yield for this community in the first and second harvest years, on average *Avicennia* leaves comprised 51.3 and 62.9 per cent, reproductive material 28.0 and 7.1 per cent, and stems 16.5 and 18.2 per cent respectively (Table 1). *Bruguiera* litter components comprised 4.2 and 11.8 per cent of the total litter yield respectively.

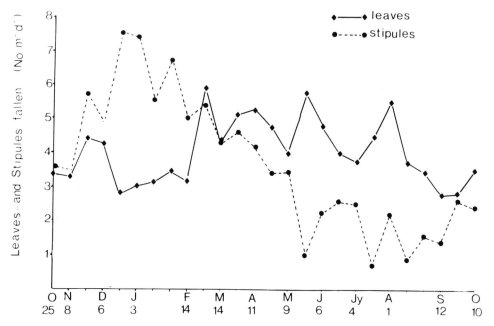

Fig. 9. Leaf and stipule fall in *Bruguiera* community in 1978/79 harvest year.

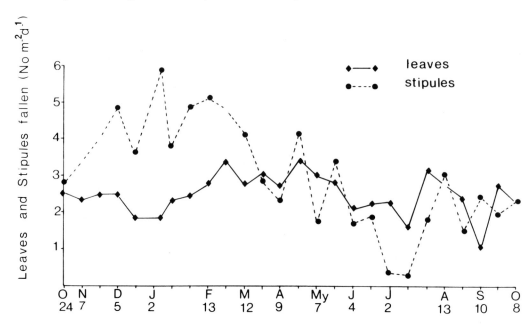

Fig. 10. Leaf and stipule fall in *Bruguiera* community in 1979/80 harvest year.

32

Bruguiera community

This community showed similar trends in total litter yields in both harvest years, although values were generally lower in the second year (Fig. 6). The total litter yield reached a peak in autumn, also largely as a result of the fall of propagules, but the increase at this time was not as marked in relation to the yields during the rest of the year as had occurred in the *Avicennia* community in 1978/79 (Figs. 7 and 8).

Furthermore, there was a more gradual rise to the high point and subsequent fall-of in the *Bruguiera* community. In this community as well, relatively high yields of litter were obtained in November/early December 1978, as explained earlier. The contribution of stem material was negligible except at specific harvests, e.g. 6 December 1978, when high winds were supposedly responsible. At the occasional harvests when relatively high yields of stem litter were recorded, this was usually due to the collection of a broken branch in the baskets. At most harvests the stem litter comprised only a small percentage of the total litter of this community and consisted only of small (usually terminal) twigs and flakes of bark (Table 1).

As there are only two trees of *A. marina* in this community, the contribution of litter by this species is negligible except at the peak of propagule abscission (Table 1).

Leaf fall varied significantly during the first year of the experiment (Figs. 7 and 9). During the warmer months, i.e. from October to March, leaf fall was relatively low, while during the cool, dry months, i.e. from April to September, leaf fall was higher. The average collections for these periods were $1.47 \, g \, m^{-2} \, d^{-1}$ and $1.86 \, g \, m^{-2} \, d^{-1}$ respectively. These results confirmed those of Lugo and Snedaker (1974) that leaf fall increased during dry periods, but do not agree with the findings of Gill and Tomlinson (1971) that peak rates of leaf fall and growth in *Rhizophora mangle* were observed during summer months. During 1979/80, in spite of a severe drought, leaf fall was constant throughout the year with values of $1.27 \, g \, m^{-2} \, d^{-1}$ and $1.25 \, g \, m^{-2} \, d^{-1}$ for the warmer and cool, dry months respectively (Figs. 8 and 10). However, in both harvest years during the warmer months rate of leaf appearance, as measured by stipule numbers, was high in comparison with the low rates achieved in the cool, dry months (Figs. 9 and 10). This suggests that in *Bruguiera* numbers of leaves per shoot do

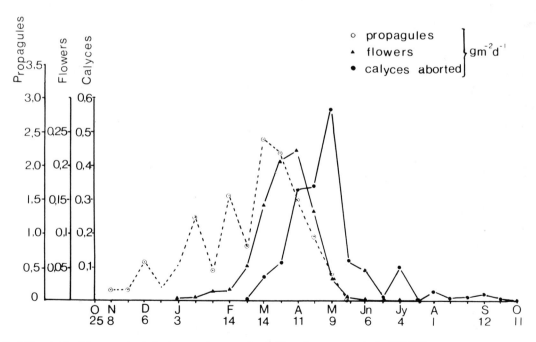

Fig. 11. Litter yields of flowers, aborted calyces and propagules in *Bruguiera* community in 1978/79 harvest year.

33

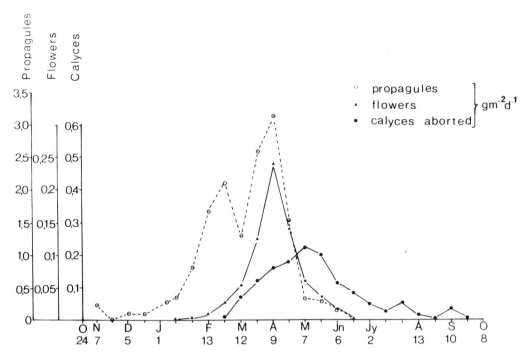

Fig. 12. Litter yields of flowers, aborted calyces and propagules in *Bruguiera* community in 1978/80 harvest year.

not remain constant, but rather that there is an increase during the warmer months. Research is in progress in an attempt to confirm this variation and also to determine the longevity of leaves produced at different times of the year.

The presence of flowers in the baskets coincided with observations on the appearance of flowers on the trees, so the beginning of January can be regarded as a fairly accurate indication of the start of flowering in *Bruguiera* (Figs. 11 and 12). There was a slow increase in flowering initially, followed by a sharp increase from mid-February, leading to a peak in early April. Thereafter there was a sharp decline in flowering which ceased completely by mid-July. The yield of aborted calyces shows a similar trend to that of flowers although collections were made up to October. The appearance of the early collections of aborted calyces indicated that many of the flowers are not pollinated successfully. While it was apparent that many of the later collections were also from unpollinated flowers, probably the very late flowers, some of the later calyces did show some, usually very slight, development of

the fruit. The reason for the abscission of the latter was not clear. It was interesting that very few buds abscised before opening. The first propagules were recorded at the beginning of November, and thereafter the yield showed an erratic increase to a maximum in March/April, followed by a steady decline until the last propagules were recorded approximately two months after the peak. With a few exceptions the early propagules were small and apparently not fully developed, although previous experience (Steinke, unpublished data) suggests that they would be capable of growing under favourable conditions.

Of the total litter yields for this community in the first and second harvest years, on average *Bruguiera* leaves comprised 64.0 and 59.5 per cent, propagules 19.3 and 27.1 per cent, calyces 5.4 and 4.9 per cent, stipules 5.0 and 4.8 per cent, flowers 1.2 and 1.3 per cent and stems 3.9 and 2.3 per cent respectively (Table 1). *Avicennia* litter components comprised only 1.2 and 1.7 per cent of the total litter yields respectively.

General discussion

From Tables 1 and 2 the mean litter production for the *Avicennia* and *Bruguiera* communities was 1.96 and 2.36 g m^{-2} d^{-1} or 7.15 and 8.61 mt ha^{-1} yr^{-1} respectively. The mean annual litter production by the mangroves of the whole Beachwood area was 231.0 mt yr^{-1} with a net primary productivity of 346.4 mt C yr^{-1}.

These figures are low by comparison with results obtained in tropical mangrove communities (Christensen, 1973; Bunt, 1979), although they compare favourably with data obtained in south Florida and Puerto Rico (Pool, Lugo & Snedaker, 1974; Teas, 1976) and in New South Wales (Goulter & Allaway, 1979). Nevertheless the productivity of the Beachwood mangroves approximates more closely to yields from temperate rather than tropical forests (Bell, Johnson & Gilmore, 1973). In spite of their relatively low productivity, it is clear that they are making a significant contribution of litter to this estuary.

In contrast to the figure of 75 to 80 per cent which represents the proportion of leaves in total litter fall in the Caribbean (Pool, Lugo & Snedaker, 1974), leaves have been shown to comprise lower percentages of the litter of these *Avicennia* and *Bruguiera* communities respectively. In this study propagules were shown to make a more significant contribution to the litter. Wood fall was low in both species, particularly in comparison with mangroves

from tropical Australia (Bunt, 1979) where wood fall formed a very much higher proportion of total litter yield. According to Pool and Lugo (1973), a high wood:litter ratio is characteristic of tropical forests, consequently the low ratio achieved in this study suggests that this behaviour is more characteristic of temperate forests.

The fact that leaf appearance is greater than leaf fall in summer and that this situation is reversed in winter suggests an adaptation to changes in radiation levels with the seasons. In summer radiation levels are high with the result that a shoot could support more leaves above their compensation point, thus favouring retention of the lower leaves which could also make a contribution towards net primary productivity. On the other hand, radiation levels are lower in winter with the result that a shoot could not support as many leaves above their compensation point. Under these conditions senescence of the lower leaves would probably be hastened. This supports the recorded observation that greater leaf fall takes place in winter, a phenomenon at least partly attributable to radiation levels. If this assumption is correct, then perhaps this seasonal variation in leaf fall is not a response which one would expect from tropical mangroves.

The fact that flowering and propagule fall followed each other closely indicates that at Beachwood a period of almost one year is required for the development of propagules after flowering. Observations in tropical Queensland suggest that under

Table 2. Mean productivity of mangrove communities at Beachwood for the years 1978/79 and 1979/80.

Community	Area ha	Mean litter yield mt ha^{-1} yr^{-1}	Total community litter mt yr^{-1}	Total NPP mt C yr^{-1}
Bruguiera gymnorrhiza	8.3	8.61	71.5	107.2
Avicennia marina	3.9	7.15	27.9	41.8
Avicennia – Bruguiera	16.7	7.88 (estimated)	131.6	197.4
Total Beachwood	28.9	–	231.0	346.4

those conditions a shorter period is required for the development of mature propagules. The longer period required at Beachwood is probably due to the fact that this species is approaching the southern limit of its distribution. Under natural conditions the southernmost known distribution of fertile *Bruguiera* trees is at the confluence of the Nxaxo and Nqusi River approximately 400 km further south (Steinke, 1972), although transplanted material, which has not yet reached a reproductive stage, is growing well as far south as East London.

It would appear, therefore, that in several respects these South African mangroves, as revealed in this study, are showing certain differences in comparison with mangrove communities in tropical areas. It is suggested that the differences in behaviour of these mangroves can be ascribed to the fact that they are growing in a subtropical region approaching the southernmost limit of their distribution.

Acknowledgements

The writers wish to thank Mr. C.J. Ward for his helpful comments and Mrs. E.L. Lawes for her assistance with the diagrams. To Messrs P. Singh and S. Kasavan thanks are due for assistance with litter collection. The co-operation of the Natal Parks, Game and Fish Preservation Board is also gratefully acknowledged. For the use of vegetation map of the Mgeni/Beachwood Nature Reserve, prepared by Miss R. Padia, we wish to express our appreciation.

Literature cited

Bell, D.T., F.L. Johnson & A.R. Filmore, 1978. Dynamics of litter fall, decomposition, and incorporation in the streamside forest ecosystem. Oikos 30:76–82.

Bunt, J.S. 1981. Studies of mangrove litter fall in tropical Australia. In: Structure, function and management of mangrove ecosystems in Autralia, ed. B.F. Clough. A.N.U. Press, Canberra.

Christensen, B. 1976. Biomass and primary production of *Rhizophora apiculata* Bl. in a mangrove in Southern Thailand. Aquatic Bot. 4:43–52.

Gill, A.M. and P.B. Tomlinson, 1971. Studies on the growth of red mangrove (*Rhizophora mangle* L.) 3. Phenology of the shoot. Biotropica 3:109–124.

Goulter, P.F.E. and W.G. Allaway, 1979. Litter fall and decomposition in a mangrove stand, *Avicennia marina* (Forsk.) Vierh., in Middle Harbour, Sydney. Aust J. Mar. Freshwater Res. 30:541–546.

Lugo, A.E. and S.C. Snedaker, 1974. Properties of a mangrove forest in southern Florida. Proc. Internat. Symp. Biol. and Mgmt Mangroves, Hawaii, pp. 170–212.

Moll, E.J., C.J. Ward, T.D. Steinke and K.H. Cooper, 1971. Our mangroves threatened. African Wildlife. 25:103–107.

Pool, D.J. and A.E. Lugo, 1973. Litter production in mangroves. In: The role of mangrove ecosystems in the maintenance of environmental quality and a high productivity of desirable fisheries, ed. S.C. Snedaker and A.E. Lugo. Rept to Bur. of Sport Fisheries and Wildlife Mgmt., USA.

Pool, D.J., A.E. Lugo and S.C. Snedaker, 1974. Litter production in mangrove forests of southern Florida and Puerto Rico. Proc. Internat. Symp. Biol and Mgmt. Mangroves, Hawaii, pp. 213–237.

Steinke, T.D. 1972. Further observations on the distribution of mangroves in the eastern Cape Province. J.S. Afr. Bot. 38:165–178.

Teas, H.J. 1976. Productivity of Biscayne Bay mangroves. In: Biscayne Bay: past, present and future, ed. A. Thorhaug, Univ. Miami Sea Grant Special Report No. 5 pp. 103–112.

CHAPTER 4

Structural features of the salt gland of *Aegiceras*

C.D. FIELD, B.G. HINWOOD and I. STEVENSON

School of Life Sciences, The New South Wales Institute of Technology. Gore Hill, Sydney, Australia

Abstract. The salt glands of *Aegiceras* have been re-examined structurally as a result of previous electro-physiological studies. It was confirmed that salt solution is secreted through a well defined lumen between the cuticular cap and the leaf cuticle, so providing a low resistance electrical pathway between the secretory cells and the external solution, by-passing the cuticular cap. It has also been shown that the secretory cell region can be considered as an equipotential region due to the presence of numerous plasmodesmata between the secretory cells.

Further evidence was found for a symplasmic continuum between the secretory cells and surrounding mesophyll cells by the demonstration of plasmodesmata connections between the basal cell of the gland, the sub-basal cells, and epidermal basal cells. The presence of chloroplasts in the sub-basal cells of the gland was also demonstrated.

Experiments using lanthanum as an electron dense tracer to define the apoplast showed evidence that this solute can reach the salt gland via the apoplast. The transport of solute to the salt glands by symplastic and apoplastic routes is, therefore, considered possible.

Introduction

The structure of the salt gland of the river mangrove *Aegiceras corniculatum* was described by Cardale and Field (1971), and it was suggested that apoplastic movement of water and solutes may occur. Figure 1 shows a vertical section of the salt gland. Campbell *et al.* (1974) showed that lanthanum could reach the salt glands of a number of halophytic plants via the apoplast and their results suggested that it might be interesting to examine *Aegiceras* in a similar manner.

Billard and Field (1973) reported that the cuticular cap, secretory cells and basal cells of the salt gland in *Aegiceras* could be distinguished according to their resting potential and voltages response characteristics. It was shown that there exists a low resistance pathway from the cuticular region to the underside of the gland. It was also demonstrated that a low resistance pathway may exist at the edge of the cuticular cap which represents a shunt from the secretory cells to the external solution. Cardale and Field (1975) have reported short circuit current experiments with leaf discs of *Aegiceras* that demonstrate that both Na^+ and Cl^- ions are actively transported by the salt glands. The present investigation was designed to re-examine the salt glands of *Aegiceras* to ascertain if there was any further structural evidence to support the physiological findings of the previous work.

H.J. Teas (ed), Physiology and management of mangroves.
© *1984 Dr W. Junk Publishers, The Hague. ISBN 90 6193 949 6. Printed in the Netherlands.*

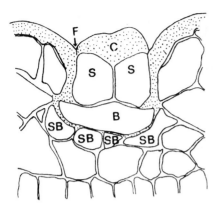

Fig. 1. Vertical section of the salt gland of *Aegiceras* showing the following structures: gland cuticle (C), junction of the gland cuticle and the inside surface of the gland (F), secretory cells (S), basal cell (B) and sub-basal cell (SB).

Materials and methods

Seedlings of *Aegiceras* were collected from the banks of the Hawkesbury River, near Sydney, Australia, and grown in soil watered with tap water. Material was selected from plants 60–90 cm in height, that were actively secreting salt.

For scanning electron microscopy fresh whole leaves or leaf strips were mounted onto stubs with double-sided adhesive tape, then rotary coated with carbon and gold palladium (60:40), and observed in a Philips PSEM 500 SEM at 12 Kv accelerating voltage. Tilt angle in all photographs was 20°.

For transmission electron microscopy, individual glands were excised and fixed in 3.5% glutaraldehyde in 0.05M sodium-cacodylate buffer (pH 7.4) for 2 hours. Glands were washed three times with cacodylate buffer, embedded in 3% agar to expedite handling and then post-fixed in 1% OsO_4 in cacodylate buffer overnight. Glands were dehydrated through an ethanol series and then flat embedded in low viscosity epoxy resin (Spurr 1969) after passing through 1:2 epoxy propane as an intermediate fluid. Resin was polymerised at 70° C for 24 hours. The flat embedded blocks were oriented to the desired plane of section and then sectioned on a Reichert OMU-2 ultramicrotone using glass knives. Sections were mounted on bare 150 or 200 mesh grids, stained with lead citrate (Reynolds 1963) and uranyl acetate (2%) and then observed in a JEM 100 U transmission electron microscope at 80 Kv accelerating voltage. Ilford EM4 plates were used for photography. Thicker sections (1 µm) were stained with 1% toluidine blue in 1% sodium borate and observed by light microscopy.

To examine the movement of lanthanum to the salt glands of *Aegiceras* a disc 2 cm in diameter was punched from the leaf, avoiding the mid-rib. The lower cuticle of the disc was gently abraded and it was then carefully washed with distilled water, dried and floated on an appropriate medium. The two solutions used were 1.2 per cent lanthanum made by raising the pH of a lanthanum nitrate solution to 7.2 by slowly adding 0.1M NaOH and 3 per cent NaCl containing 1.2 per cent lanthanum nitrate (pH 7.2). Distilled water and 3 per cent NaCl solution were used as controls. The leaf discs were treated for 24 hours under continuous illumination. The upper surface of the discs used was examined microscopically for the presence of excreted salt. The material was then treated as described above for transmission electron microscopy except that it was viewed unstained. The method used was similar to that described by Campbell *et al.* (1974).

Results

Scanning electron microscopy

In *Aegiceras* active and non-active salt glands are found on both the upper and lower surface of the leaves, but no stomata are found on the upper surface of the leaves. Figure 2 shows glands with two differing surface morphologies on the upper surface of a mature leaf. One of the salt glands shown exhibits a raised rim while the other one has a more crater-like appearance. The gland with the raised rim contains a thin layer of crystallised salt around its inner surface. No plug of cuticular material is evident in the top of the gland. The other salt gland shows no evidence of salt crystals but the cuticular plug is clearly present. The visible junction between the central cuticular plug and the inner surface of the gland should be noted. Figure 3

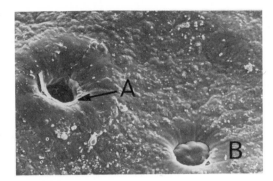

Fig. 2. The upper surface of a mature leaf, showing salt glands with differing surface morphologies, a raised rim (A) and a more creater-like form (B). Note salt secretion on the side of the crater surrounding (A) (arrow). ×320.

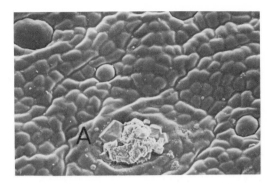

Fig. 3. The upper surface of a juvenile leaf showing salt glands of differing sizes, including an active salt encrusted gland (A). The leaf cuticle is much more wrinkled than that of the mature leaf. ×320.

shows a similar view of the upper surface of a juvenile leaf, the size of the glands varies considerably and the surface is disinctly more wrinkled than in the case of the mature leaf. In this instance the glands are more closely packed, but the central cuticular plug is evident and one of the glands is actively excreting salt. Active glands on the surface of juvenile leaves were frequently observed but differences in gland external size did not appear to be related to the ability to secrete salt. An actively excreting salt gland from a mature leaf is shown in Figure 4. The complete peripheral layer of crystallised salt was regularly observed and appeared to be characteristic of active salt glands. In the particular example shown the presence of a depressed

central cuticular plug can be seen in a typically crater-type gland. A salt gland from a juvenile leaf is shown in Figure 5. The cap of the central cuticular plug dominates the centre of gland and the lumen of the junction of the cuticular core and the inside surface of gland through which the salt secretion appears to emerge is very marked.

Transmission electron microscopy

Figure 6 shows a general view of the parts of three secretory cells in transverse section. The cells are similar to those previously reported by Cardale and Field (1971) in that they contain dense cytoplasm with no large central vacuole and smooth cell walls

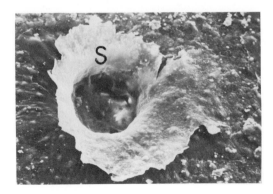

Fig. 4. An actively secreting salt gland from a mature leaf showing a peripheral ring of salt secretion (S) that was regularly observed. ×640.

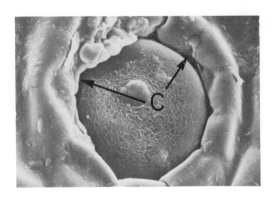

Fig. 5. A salt gland from a juvenile leaf. Note the cap of cuticular material at the top of the gland (C) and the gap between cuticular plug and inner gland surface through which the secretion emerges (arrows). ×1250.

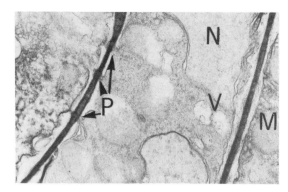

Fig. 6. Portion of three secretory cells in transverse section. Note nucleus (N), mitochondria (M), and vesicles (V). Cell to cell contacts through plasmodesmata (P) (arrows) are evident. ×27000.

separating them. The interesting difference is the presence of well defined plasmodesmata that form frequent cell to cell links. There is also confirming evidence (Fig. 7) of plasmodesmatae connections between the basal cell of the gland and the sub-basal cells, a group of small cells immediately below the basal cell. Both the basal cell and the sub-basal cells are vacuolate. The sub-basal cells were found to contain chloroplasts similar to those found in the mesophyll palisade cells (Fig. 8) whereas no evidence was found of chloroplasts or plastid-like organelles in the basal or secretory cells. It was also found that a small number of plasmodesmata link the sub-basal cells to the layer of cells immediately below them, which can be referred to as the epidermal basal cells (Fig. 8). Three layers of epidermal basal cells occur between the sub-basal and the mesophyll cells. All these cells were observed to be linked by plasmodesmata. The epidermal basal cells are vacuolate with a small peripheral

cytoplasm that does not appear to contain any plastids.

The leaf discs of *Aegiceras* floated on 3% NaCl solution exhibited normal salt secretion but those floated on lanthanum showed a marked inhibition of salt secretion. After exposure of the leaf tissue to lanthanum electron-dense deposits were observed throughout the apoplast of the leaf. They were heavy in the walls of the mesophyll cells. There were also deposits at the junction of the basal cell and the secretory cells and in the inner cell wall material of the salt glands (Fig. 9). No deposits resembling those seen in the discs treated with lanthanum were found in the walls of the discs floated on NaCl or distilled water (Fig. 10).

Discussion

The structure of the salt gland of *Aegiceras* has previously been described by Cardale and Field (1971). The present studies have been designed to extend that work and to examine the structure of the gland for evidence to support the electrical

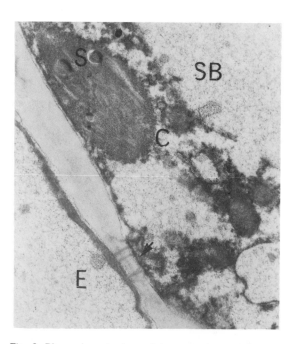

Fig. 8. Plasmodesmata (arrow) between the vacuolate sub-basal cell (SB) and an epidermal basal cell (E). Note chloroplast (C) with starch grains (S). ×50000.

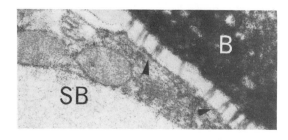

Fig. 7. Plasmodesmata (arrows) between the basal cell (B) and the sub-basal cell (SB). ×50000.

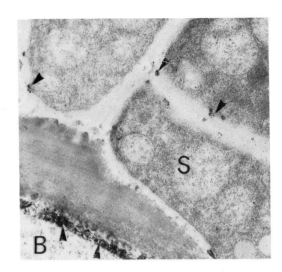

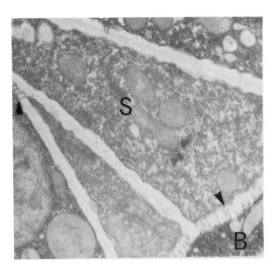

Fig. 9. Unstained La⁺ nitrate – NaCl treated salt gland, showing heavy metal deposits in walls between secretory cells (S) and in the wall of the basal cell (B) (arrows). ×20400.

Fig. 10. Unstained NaCl treated salt gland, showing plasmodesmata between adjacent secretory cells (S), and between secretory and basal cells (B) (arrows). The walls between secretory cells show no heavy metal deposits. ×14000.

properties of the gland reported by Billard and Field (1974). These earlier studies revealed no evidence of pores in the gland cuticle and suggested that the cuticular cap was a region of high electrical resistance. The present results confirm these findings and indicate that the salt solution is secreted through a peripheral lumen between the cuticular cap and the leaf cuticle. This is in marked contrast to *Aegealetis, Limonium* and *Tamarix* (Hill and Hill, 1973), which all possess cuticularized pores through which the secretion is extruded. The results also give excellent support to the electrical finding that a pathway exists at the upper boundary of the gland and the leaf cuticle that represents an electrical shunt from the secretory cells to the external solution by-passing the cuticle phase. The presence of a well defined lumen at the periphery of the gland also supports the previous finding that in *Aegiceras* the transport of salt through the cap is almost entirely in association with bulk fluid movement. It is interesting to note that in juvenile leaves the glands that secrete salt appear to have a similar surface appearance and size to glands that appear inactive. This suggests that not all glands become active simultaneously and in fact, some may remain inactive for a long period of time.

The previous electrical studies indicated that the

secretory cells should be connected by low resistance pathways, so that the region could be represented as being equipotential. The demonstration that cell to cell plasmodesmata commonly exist supports the concept of an equipotential region and helps validate the use of a microelectrode technique to investigate electrical properties of a multicellular gland. The presence of plasmodesmata between the basal cell of the gland and the sub-basal cells extends the evidence for a symplasmic continuum between the secretory cells and surrounding mesophyll cells. This would appear to have similarities with other salt gland systems (Lüttge, 1975). However, whether the presence of numerous plasmodesmata found between the salt gland cells and peripheral cells means that this is the only route for ion movement, as suggested by Lüttge (1975), is doubtful in the case of *Aegiceras*. The experiments using lanthanum suggest that solute can reach the glands without entering the mesophyll cells of the leaf and that apoplastic transport into the gland may be possible. This finding supports the work of Campbell *et al.* (1974). It is also interesting to note that Billard and Field (1974) suggested that a low resistance electrical pathway exists from the cuticular region to the underside of the gland, which shunts the secretory cells. This low resistance electrical pathway could be accounted for if there was apoplastic transport of sodium chloride solution into the gland. Its presence gives additional support for at least limited apoplastic transport of solute into the gland.

Cardale and Field (1971) could find no evidence of chloroplasts in the secretory cells or basal cell and the present results are in agreement with that finding. The presence of chloroplasts in the sub-basal cells as shown in Figure 8 suggests the involvement of photosynthetic energy transfer reactions in the movement of ions into the base of the gland. This finding supports the suggestion of Billard and Field (1974) that the charge separation which can be observed across the whole leaf is associated with a large light dependent resting potential gradient generated at the base of the gland.

The results presented in this paper give good anatomical and cytological support for the electrical properties of the salt glands that have been suggested previously.

Acknowledgements

C.D. Field wishes to acknowledge financial support from the Australian Research Grants Committee.

Literature cited

Billard, B. and C.D. Field. 1974. Electrical properties of salt gland of *Aegiceras*. Planta (Berl.) 115:285–296.

Campbell, N., W.W. Thomson and K. Platt. 1974. The apoplastic pathway of transport to salt glands. J. Exp. Bot. 25:61–69.

Cardale, S. and C.D. Field. 1971. The structure of the salt gland of *Aegiceras corniculatum*. Planta (Berl.) 99:183–191.

Cardale, S. and C.D. Field. 1975. Ion transport in the salt gland of *Aegiceras*. Proceedings of the International Symposium on Biology and Management of Mangroves, Vol. II, p. 608–614. University of Florida.

Hill, A.E. and B.S. Hill. 1973. The *Limonium* salt gland: a biophysical and structural study. Int. Rev. Cytol. 35:299–319.

Lüttge, U. 1975. Salt glands. In Ion transport in plant cells and tissues, p. 335–376, Baker, D.A. and Hall, J.L. (Eds). North Holland.

Reynolds, E.S. 1963. The use of lead citrate at high pH as an electron opaque stain in electron microscopy. J. Cell Biol. 17:208–212.

Spurr, A.R. 1969. A low-viscosity epoxy resin embedding medium for electron microscopy. J. Ultrastruct. Res. 26:31–43.

CHAPTER 5

Ions in mangroves

C.D. FIELD

School of Life Sciences, The New South Wales Institute of Technology, Gore Hill, Sydney, Australia

Introduction

Halophytes are plants which complete their life cycle in an environment which has a high salt content and may be contrasted with glycophytes which are plants that are intolerant of such high salt levels. Mangroves are classed as halophytes as they are trees and shrubs that grow primarily in saline habitats. The definition is vague in two important aspects, namely the nature of the salts involved and their concentrations. For species such as *Rhizophora stylosa* which grows in seawater, the concentration of salt water content is set by the composition of sea water (Table 1). Higher concentrations will naturally occur during periods of high insolation between tides and lower concentrations during periods of rainfall. Other species, such as *Aegiceras corniculatum*, appear to prefer less saline habitats. Table 2 shows some data on the effect of salinity on growth of five species of mangrove. It is clear that there are some discrepancies in this data but that for these species at least there is some evidence for salt mediated growth. This does not mean that these mangroves require salt for successful growth and most mangroves probably grow reasonably well in freshwater. Indeed, it has been suggested by Snedaker (1979) that freshwater is a physiological requirement and salt water is an ecological requirement. The former prevents excess respiratory losses and the latter prevents invasion and competition of non-halophytes. It may be reasonable therefore to assume that most mangroves are not obligate halophytes.

However, all mangroves in their normal habitat maintain water within their symplast against a considerable osmotic gradient, since the cellular water potential must be more negative than that of the surroundings if the plant is to maintain its water content. This paper will be concerned with the mechanism of the osmotic adjustment, its relation to ion uptake and with the ability of mangroves to accumulate potassium from an environment dominated by sodium ions. Consideration of the wider aspects of the uptake of other essential ions which are present only in relatively low concentrations is precluded by a lack of data.

Table 1. The major constituents of an ocean water after Harvey (1966).

Element	Concentration		% Weight of ions
	gKg^{-1}	milli molar	
Sodium	10.8	483.	30.7
Magnesium	1.3	54.5	3.7
Calcium	0.41	11.5	1.2
Potassium	0.39	10.	1.1
Strontium	0.01	0.1	.02
Chloride	19.4	558.	55.3
Sulphate	2.7	29.	7.7
Bromide	0.07	0.4	.2
Borate	0.03	0.5	.08

H.J. Teas (ed), Physiology and management of mangroves.

Table 2. Effect of salinity on growth of mangroves.

Species	Concentration of NaCl for maximum growth	Reference
Avicennia marina	20% seawater	Clarke and Hannon (1970)
Aegiceras corniculatum	20% seawater	Clarke and Hannon (1970)
Avicennia marina	50% seawater	Connor (1969)
Rhizophora mangle	100% seawater	Stern and Voigt (1959)
Rhizophora mangle	25% seawater	Pannier (1959)

Osmotic adjustment

The osmotic potential of ocean water is of the order of −2300 KPa. This value might therefore be expected to set the upper limit of the water potential for marine halophytes. Walter and Steiner (1936) showed that the expressed sap of leaves and roots of various mangrove trees exceeded those of seawater, with those of the leaves having the highest potential. Table 3 shows some examples of the values obtained for leaves.

There is general uniformity in the values obtained except for *Avicennia marina,* which has a greater range. Such a gradient of osmotic potential between the external medium and the sap could have explained the entry of water from the sea into the plants.

Except in the case of those mangroves with salt glands in the leaves there is very little evidence of progressive salt accumulation in the plant, and so it

can be hypothesised that the transpiration stream must be low in concentration of salt. This proposition was first advanced by Walter and Steiner (1936) and apparently confirmed by Scholander *et al.* (1964, 1965, 1966 and 1968) and Gessner (1967).

The values obtained by Scholander for the osmotic potential of the xylem sap for various species of mangrove are shown in Table 4. It is clear that these potentials are well above the osmotic potential of seawater and that therefore water movement would not take place on a purely osmotic basis. It is interesting to note that Scholander only measured the osmotic potential of the three mangroves shown in Table 4. In other mangroves studies the chloride concentration in the xylem sap was determined.

There appears to be only one value in the literature on the Na^+, K^+, and Cl^- concentrations in the xylem sap of a mangrove. Atkinson *et al.* (1966) give a value for *Aegialitis* of 122 mM Cl^-, 118 mM Na^+ and 14 mM K^+. There is good evidence from Joshi (1975) (Table 5) that there must be an effective uptake mechanism for K^+ ions in most mangroves. It is interesting to speculate that in some species the K^+ ion may dominate the Na^+ ion in the xylem sap. For instance in the case of *Rhizophora,* Scholander states a figure of 20 to 26 mM Cl^- in xylem sap, and Atkinson gives a figure of 17 mM. If the potassium ion concentration in the sap was say twice that in seawater, that is 20 mM, it could account for most of the necessary cations, rather than sodium. There is clearly a need to re-examine the ion content of the xylem sap of many of the mangroves to ascertain the true relationship between the sodium and potassium ions.

Table 3. The osmotic potential of the expressed sap from the leaves of some mangroves (Walter and Steiner 1936)

Species	Osmotic Potetial (KPa)
Sonneratia alba	− 3290 to − 3550
Rhizophora mucronata	− 3530 to − 3590
Ceriops tagal	− 2820 to − 3670
Avicennia marina	− 3450 to − 6200
Bruguiera gymnorrhiza	− 3340
Lumnitzera racemosa	− 2860 to − 3620

Table 4. The osmotic potential of the expressed xylem sap of some mangroves (Scholander 1966).

Species	Osmotic potential (KPa)
Avicennia nitida	− 300 to − 600
Rhizophora mangle	− 50 to − 150
Laguncularia racemosa	− 120 to − 150

Table 5. The ion contents of a number of mangroves.

Species	Ion Content (mM $\cdot$ gDW^{-1})						Reference
(Plant leaves)	Ca^{2+}	Mg^{2+}	K$^+$	Na$^+$	Cl$^-$	Na$^+$/K$^+$	
Rhizophora mucronata	–	–	0.11	0.84	1.02	7.6	Atkinson (1976)
Aegialitis annulata	–	0.49	0.20	0.62	0.63	3.1	Atkinson (1967)
Rhizophora mucronata	0.28	0.32	0.11	1.46	2.14	13.3	Joshi (1975)
Bruguiera gymnorrhiza	0.14	0.37	0.06	1.50	1.42	25.0	Joshi (1975)
Avicennia officinalis	0.18	0.47	0.32	0.86	0.42	2.7	Joshi (1975)
Aegiceras majus	0.10	0.13	0.23	1.02	1.03	4.4	Joshi (1975)
Seawater (mM)	11.5	54.5	10.0	483	558	48.3	

The scholander concept of water relations in mangroves

The current concept of water relations in mangroves is that proposed by Scholander (1965) and shown in Fig. 1. The water potential is given by the sum of the osmotic potential and the hydrostatic potential.

In seawater it is assumed that the hydrostatic potential is zero and the osmotic potential −3000 KPa. In the vacuole of the root cell which is considered turgid, the hydrostatic pressure is considered to be 1000 KPa and the osmotic potential −4000 KPa. The experimental finding that the osmotic potential of the xylem sap is close to zero means that a hydrostatic pressure potential of at least −3000 KPa must exist in the xylem vessels for water to flow. Scholander postulates that this negative hydrostatic pressure potential is generated by an equivalent osmotic potential in the cells of the leaves. If there is evaporation from the cell the balance is upset and the potential drop is transmitted through the system to the seawater. The resultant transpiration stream is purely diffusive through the leaf and root membrane and depends upon bulk flow through the xylem. Freshwater is pulled from the seawater through a nearly balanced system by a process of ultra filtration.

Scholander pointed out that if the stem were cut and pressure applied to the leaves which was just sufficient to cause sap to appear at the cut end, then that pressure should be equivalent to the osmotic potential of the leaf cells.

Table 6 shows some values that were obtained by Scholander (1965 and 1968) and it is clear that in all cases the difference in water potential is sufficient to produce a transpiration stream. It is perhaps interesting to note that there is very little difference between species even though it has been found that they have marked differences in the way they handle their NaCl content.

An analysis of Scholander's model shows that the cell membranes in both the root and the leaf must be able to withstand an hydrostatic gradient of at least 4000 KPa (40 ATMS) which is a very considerable pressure gradient. The model also assumes that the properties of the cell membrane in the root are such that ultrafiltration of ions can occur and that the osmotic potential in the leaf cells is less than −4000 KPa. The model also assumed that there is a rigid stem provided with capillary channels so that water under tension will not cavitate. There is very little experimental data to sup-

Table 6. Sap tensions in the xylem of various mangroves (Scholander 1965, 1968).

Species	Applied Pressure (KPa)	Excess over Seawater (2400 KPa)
Rhizophora mangle	3800–5200	1400–2800
Avicennia marina	3800–5400	1400–3000
Bruguiera gymnorrhiza	3700–4000	1000–1600
Ceriops tagal	3000–5000	600–2600
Aegiceras corniculatum	3800–5100	1400–2700
Aegialitis annulata	3100–5300	700–2900

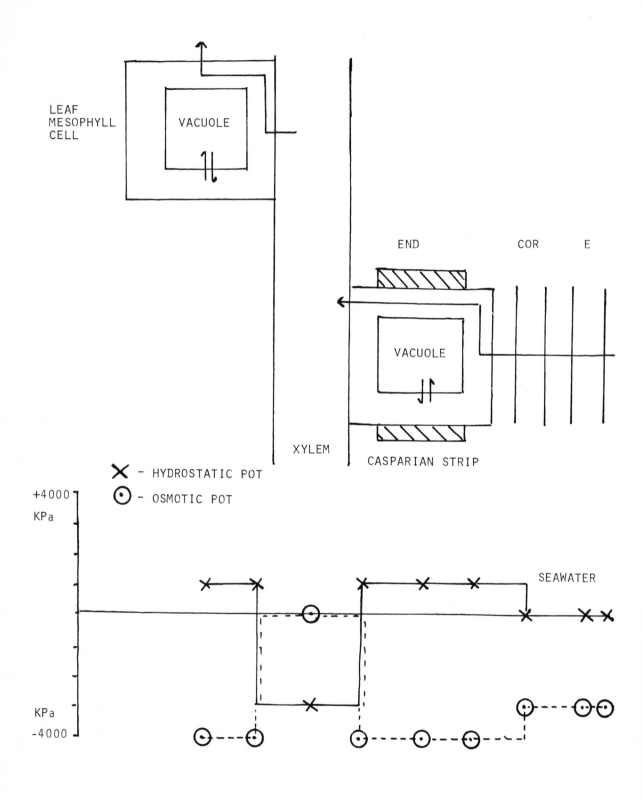

Fig. 1. The variation of the osmotic potential and the hydrostatic potential in a mangrove plant immersed in seawater (after Scholander 1965).

port many of these assumptions and there is a need to design experiments to give further empirical support for Scholander's hypothesis. Very recently Tyree (1976) has raised the question of the existence of negative pressure (tension), and Andrews (1976) argues that it is very unlikely that tension states occur in the xylem of trees. Regardless of the mechanisms by which water is transported between the root and the leaves of a plant the chemical potential difference of water between root and atmosphere represents the overall driving force for the ascent of water in plants. It is pertinant to examine one or two aspects of the model. First it must be considered whether salt accumulation in the leaves can explain the lowering of the plant water potential by the lowering of the solute potential. It has been shown that sap solute potentials in mangroves as measured by the 'pressure bomb' technique are in the range 2800–6200 KPa. There are, however, few data on the distribution of ion concentrations in the leaves of mangroves. It should be noted that the solute potential calculated from ion concentrations compared with that from the pressure bomb are not always in agreement (Wallace and Kleinkopf, 1974).

The vacuole comprises some 95% of the mature leaf surface volume and it is a commonly held view that the ions accumulated for osmotic adjustment are localised within it. It is therefore possible to estimate that the concentration within this compartment is close to that which can be calculated from the tissue as a whole, that is 500–700 mM. This figure would be sufficient to satisfy the Scholander model. However, in those mangroves with glands or those that exhibit succulence it is conceivable that a very non-uniform distribution of NaCl concentration exists in the leaf. Indeed there is some evidence that the salt in the xylem can pass to the salt glands without equilibrating with the main chloride pool of the leaf (Atkinson, 1967; Teas, 1979). It is dangerous to try and evaluate average values of the components of the water potential. The Scholander model may therefore be far too simple when applied to the complex leaf tissue that exists in many mangroves.

It is interesting also to observe that there is no evidence in mangroves as to the distribution of ions between the vacuole and the cytoplasm in the leaf or the root. Since some mangroves show a positive growth response to NaCl it is possible that such plants are adapted metabolically to function at high ion concentrations. There is no direct experimental evidence on this proposition but evidence from studies on other halophytes (Flowers, 1976) suggests that there are no salt mediated enzymes present and that the NaCl in the cytoplasm may be low due to the presence of high levels of amino acids in the cytoplasm. The other assumption to consider is that ultrafiltration of seawater occurs at the roots.

Scholander (1968) showed evidence that when a pressure of 4000–4500 KPa was applied to a detopped Avicennia seedling root in a bathing solution containing 651 mM NaCl (somewhat higher than normal seawater) that an exudate was obtained from the root containing 34 mM NaCl (somewhat lower than other experimental data on sap, which give values of 68 to 136 mM NaCl). Scholander pointed out that the roots apparently rejected some 95% of NaCl in the bathing solution. Attempts to repeat Scholander's experiment on seedlings of Avicennia have failed to produce similar values. Results reported elsewhere in this publication show that the roots of Avicennia seedlings will only reject 30% of the NaCl in seawater. In addition, the hydraulic conductivity of the roots of Avicennia seedlings have been shown to be similar to that for other plant species.

These results are in serious disagreement with the proposed model of Scholander. It is interesting, however, to consider the data given by Chapman (1944) for the % NaCl in seedlings of Avicennia germinans (Table 7). It is evident that once a seedling becomes rooted that there is a sharp increase in NaCl content, and that this is followed by a fall to values found in adult tissues. The seedlings must

Table 7. % NaCl in developing *Avicennia germinans* (after Chapman 1944)

Plant Material	Seeds	Seedlings	Mature Plants
Stem	–	10	3
Root	3	13	5
Leaf	–	30.5	5

have a remarkable capacity to absorb salt without damage to the tissues, but it would seem that subsequently some form of regulatory mechanism must become involved to keep the salt concentration at an acceptable value. It is conceivable that in the experiments referred to previously where the roots only rejected 30% of the NaCl in seawater that the seedlings of *Avicennia* studied, which were usually 20 cm in height with two to four leaves, had not yet developed a root system capable of ultrafiltering seawater.

A further variable for which there is no data on mangroves is the elastic coefficient, ε, of the cell walls. Dainty (1976) has shown that the elastic coefficient is an important parameter in cell water relations. The rate of water exchange during shrinking or swelling of a plant cell in response to external osmotic stress is not only controlled by the properties of the membrances but also by the elasticity of the cell wall.

Conclusion

It appears from the literature that mangroves regulate their ion content by a combination of four different methods: salt exclusion, salt secretion, succulence or discarding salt-laden organs. The physical and biochemical basis of these methods is far from understood and a much more precise experimental description of the physiological processes in mangroves is required before a satisfactory understanding of the mechanisms involved is achieved.

Literature cited

Atkinson, M.R., G.P. Findlay, A.B. Hope, M.G. Pitman, H.D.W. Saddler and H.R. West. 1967. Salt regulation in the mangroves *Rhizophora mucronata* Lam and *Aegialitis annulata* R. Aust. J. Biol. Sci. 20:589–599.

Chapman, V.J. 1944. 1939 Cambridge University Expedition to Jamaica. I. A study of the botanical processes concerned in the development of the Jamaican shore-line. J. Linn. Soc. London Bot. 52:407–447.

Clarke, L.D. and N.J. Hannon. 1970. The mangrove swamp and salt marsh communities of the Sydney district. III. Plant growth in relation to salinity and waterlogging. J. Ecol. 58:351–369.

Connor, D.J. 1969. Growth of grey mangrove (*Avicennia Marina*) in nutrient culture. Biotropica. 1:36–40.

Dainty, J. 1976. Water relations of plant cells. In Encyclopedia of Plant Physiology. Vol. 2.A. Ed. U. Luttge and M.G. Pitman pp. 12–35. Springer-Verlag.

Flowers, T.J., P.F. Troke and A.R. Yeo. 1977. The mechanism of salt tolerance in halophytes. Ann. Rev. Plant Physiol. 28:89–121.

Gessner, F. 1967. Untersuchungen an der mangrove in Ost-Venezuela. Int. Res. Ges. Hydrobiol. 52:769–781.

Hammel, H.T. 1976. Colligative properties of a solution. Science. 192:748–755.

Harvey, H.W. 1966. The Chemistry and Fertility of Sea Water. pp. 36–54. Cambridge University Press.

Joshi, G.V., B.B. Jamale and L.J. Bhosale. 1975. Ion regulation in mangroves. In Proceedings of International Symposium on Biology and Management of Mangroves. Eds. G. Walsh. S. Snedaker and H. Teas pp. 595–607. University of Florida.

Pannier, P.F. 1959. El efecto de distintas concentrationes sotenas sobre el desarrolo de *Rhizophora mangle*. Acta Cient. Venz. 10:68–76.

Scholander, P.F., H.T. Hammel, E.A. Hemmingsen and E.D. Bradstreet. 1964. Hydrostatic pressure and osmotic potential in leaves of mangroves and some other plants. Proc. Nat. Acad. Sci. 52:119–125.

Scholander, P.F., H.T. Hammel, E.D. Bradstreet and E.A. Hemmingsen. 1965. Sap pressure in vascular plants. Science. 148:339–346.

Scholander, P.F., E.D. Bradstreet, H.T. Hammel and E.A. Hemmingsen. 1966. Sap concentrations in halophytes and some other plants. Plant Physiol. 41:529–532.

Scholander, P.F. 1968. How mangroves desalinate seawater. Physiol. Plant. 21:258–268.

Snedaker, S.C. 1979. Australian institute of Marine Science. National Mangrove Workshop, Townsville, Australia. 18–20 April. In press.

Stern, W.L. and G.K. Voight. 1959. Effect of salt concentration on growth of red mangrove in culture. Bot. Gaz. 121:36–39.

Teas, H.J. 1979. Silviculture with saline water. In The Biosaline Concept. Ed. Alexander Hollaender pp. 117–161. Plenum press.

Tyree, M.T. 1976. Negative turgor pressure in plant cells: fact or fallacy. Can. J. Bot. 54:2738–2746.

Wallace, A. and G.E. Kleinkopf. 1974. Contribution of salts to the water potential of woody plants. Plant Sci. Letters. 3:251–257.

Walter, H. and M. Steiner. 1936. Die Okologie der ost-afrikanischen mangroves. Z. Bot. 30:65–193.

CHAPTER 6

Movement of ions and water into the xylem sap of tropical mangroves

C.D. FIELD

School of Life Sciences, The New South Wales Institute of Technology, Gore Hill, Sydney, Australia

Introduction

The physiology of the roots of mangroves has received relatively little attention. The present experiments are concerned with the mangrove *Avicennia marina* and the accepted behaviour of this mangrove was described by Scholander (1962, 1966). It is believed that this mangrove only partially excludes salt at its roots but that it is capable of excreting large quantities of salt from its leaves. The xylem sap concentration for *Avicennia* is stated as being 10–20% of the NaCl present in seawater.

Chapman (1944) stated that once a seedling of *Avicennia* becomes rooted there is a sharp increase in Cl⁻ content and that this is followed by a fall to the values found in adult tissues. The distribution of NaCl between root, stem and leaves was 1:.8:2.4 when the level of salt in the plant was at a maximum. It was argued that seedlings must have a remarkable capacity to absorb salt without damage to tissues.

The present study examines the hydraulic conductance of roots of *Avicennia marina* seedlings and determines the reflection coefficient for Na⁺ and Cl⁻ ions.

Theory

The model used was that of a simple two compartment system consisting of an outer compartment, root bathing solution and an inner one representing the xylem exudate. Everything separating these two compartments is treated as a single membrane. The transport equation that is applied is:

$$J_v = L_p (\Delta p - \sigma \Delta \Pi)$$

where J_v is total volume flow in cm sec^{-1}, Lp is the hydraulic conductivity in cm sec^{-1}KPa^{-1}, Δp is the pressure difference across the root system in KPa, σ is the reflection coefficient and $\Delta \Pi$ is the osmotic pressure difference between the bathing solution and the xylem exudate in KPa.

This equation predicts that the slope of the graph J_v against Δp will give the value of Lp, if the term $\sigma \Delta \Pi$ becomes small relative to Δp. The value of σ is obtained from a knowledge of the limiting values of ion concentrations in the exudate and the ion concentration in the bathing solution.

Method

Seedlings of *Avicennia marina* were collected from the Sugar Loaf Bay region of Sydney Harbour, Australia. The seedlings were 15 to 20 cm in height with two or four leaves. They were initially grown in a greenhouse in pots with a mixture of mud and compost and watered with either 100% seawater or 10% aquasol solution. After a month when the seedlings had become established they were removed from the pots by carefully washing the roots and then placing them in unaerated flasks with the appropriate solution. All solutions were changed on a weekly basis. Seedlings were only used if they had been grown in solution for more than three months and if they were apparently healthy and excreting salt from the leaves.

The experimental procedure was essentially the same as that used by Fisus (1977). The stem of the seedling was severed a few centimeters above the root system. The root system was in an unaerated solution which was either 100% seawater or a 10% aquasol solution. The stem was sealed into the lid of a pressure vessel, that was in turn bolted to the pressure vessel. The exudation rate from the stem was determined at various levels of applied pressure by measuring the rise of xylem exudate up a calibrated thin glass tube using a travelling microscope.

The ion concentrations in the exudate and the bathing solution were measured using a Varian atomic absorption spectrometer and Radiometer Chloride meter. Total root surface area was estimated at the end of the experiment using a modification of the method described by Newman (1966).

Results

The data from a typical volume flow-pressure experiment is shown in Fig. 1. The graph shows that there is zero flow at zero pressure, which confirms that there is no root exudate without applied pressure. The variation of volume flow with applied pressure is initially non-linear but eventually a linear region is obtained. The hydraulic conductivity (L_p) of the root is given by the slope of the line at high flow rates, when the system is well into the linear operating position of the curve.

The values found for L_p from five experiments for roots of *Avicennia* seedlings are shown in Table 1. It should be noted that a simple Student 't' test shows that the two values are significantly different at a 95% confidence level.

The average values for five separate experiments of the reflection coefficient for Na^+ and Cl^- using roots of *Avicennia* seedlings grown in 100% seawa-

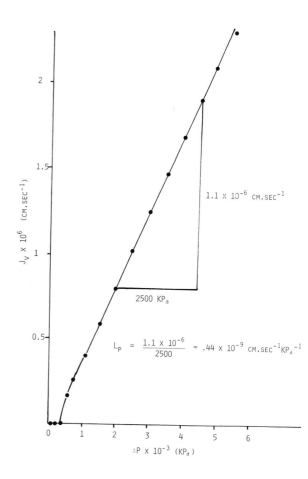

Fig. 1. Volume flow in an *Avicennia* root system in seawater as a function of applied pressure

Table 1. Hydraulic conductivity of *Avicennia marina* roots.

Roots in 100% Seawater	$L_p = .33 \pm SD.11 \times 10^{-9}$ cm $\cdot$ sec^{-1} KPa^{-1}
Roots in 10% Aquasol Solution	$L_p = .75 \pm SD.21 \times 10^{-9}$ cm $\cdot$ sec^{-1} KPa^{-1}

(Each result is the average of five separate experiments)

ter are shown in Table 2. In every experiment the value of the K^+ ion concentration in the exudate exceeded that of the seawater. In the experiments using 10% aquasol solution the values of the Na^+, K^+, and Cl^- ion concentration in the exudate always exceeded that of the solution. Table 3 shows the concentrations of various ions in the pressed sap from separate parts of the plants grown under different conditions.

50

Discussion

It is interesting to compare the value of hydraulic conductivity (Lp) measured for *Avicennia* roots with values measured for other plants. Table 4 shows the ranges of values that have been obtained for various plants. The values stated are not entirely consistent but the values obtained for *Avicennia* fall at the bottom end of the range.

It is also interesting to note that the value for Lp for *Avicennia* is comparable to those found for organisms found in seawater and to cellulose acetate, which is an artificial membrane used in desalination by reverse osmosis (Motais, 1969; Thau, 1966). The apparent difference found for Lp between *Avicennia* roots grown in seawater and those grown in 10% aquasol may mean that changes occur in the roots of the mangrove as a result of the level of salinity that it is exposed to while growing.

Table 2. Reflection coefficients for Na^+ and Cl^- ions in *Avicennia marina* roots in 100% seawater.

Na^+ ions	$\sigma = .31 \pm SD.09$
Cl^- ions	$\sigma = .34 \pm SD.09$

(Each result is the average of five separate experiments)

Table 3. Ion concentrations in seedlings of *Avicennia marina*.

Growth Medium	Plant Material	Ion Concentration of Pressed Sap (mM) [$\pm$ SD]		
		Na^+	K^+	$Cl-$
Seawater	Stem Xylem	792 ± 173	118 ± 25	799 ± 178
[Na^+ 510mM	Root Xylem	737 ± 144	228 ± 34	916 ± 156
K^+ 13.0mM				
Cl^- 576mM]	Fine Roots	527 ± 72	198 ± 64	667 ± 77
10% Aquasol solution	Stem Xylem	241 ± 43	72 ± 9	239 ± 46
[Na^+ 2mM	Root Xylem	205 ± 77	158 ± 73	297 ± 115
K^+ .8mM				
Cl^- 1mM]	Fine Roots	153 ± 87	182 ± 48	225 ± 89

(All values are the average of five experiments)

Table 4. Hydraulic conductivity for higher plant cells.

Plant Species	$L_p \times 10^9$ $(cm \cdot sec^{-1} KPa^{-1})$	Reference
Maize Onion Broadbean Sunflower Tomato	$0.5 \leq 10$	Newman (1973)
Soy Bean	26.5	Fiscus (1977)
Tomato	10	Shalhevet (1976)
Sunflower	4.6	Shalhevet (1976)
Broadbean	$25 \leq 50$	Brouwer (1954) (as analyzed by Fiscus 1977)
Avicennia (seawater)	.33	Field
Avicennia (aquasol)	.75	Field

The value for the reflection coefficient for sodium and chloride ions implies that the root does not reject these ions with much efficiency. It is interesting to note that the results stated by Scholander (1968) give a result of $\sigma = .95$. In the present experiments it was found impossible to reproduce the results of Scholander as the exudate was always at least 50% of the salinity of the seawater bathing the roots. It may be that in seedlings of *Avicennia* at least much greater quantities of salt find their way through the xylem and tissue of the plant than is normally considered to be the case. The suggestion is partially supported by the data obtained for the ion concentrations found in expressed sap for various parts of the seedling of *Avicennia*.

Literature cited

Brouwer, R. 1954. The regulating influence of transpiration and suction tension on the water and salt uptake by roots of intact *Vicia Faba* plants. Acta bot. neerl. 3:264–312.

Chapman, V.J. 1944. 1939 Cambridge University Expedition to Jamaica. I. A study of the botanical processes concerned in the development of the Jamaican shore-line. J. Linn. Soc. London Bot. 52:407–487.

Fisus, E.L. 1977. Determination of hydraulic and osmotic properties of Soybean root systems. Plant Physiol. 59:1013–1020.

Motais, R., J. Isaia, J.C. Rankin and J. Maetz. 1969. Adaptive changes of the water permeability of the teleostean gill epithelium in relation to external salinity. 51:529–546.

Newman, E.I. 1966. A method of estimating the total length of root in a sample. J. Appl. Ecol. 3:139–145.

Newman, E.I. 1973. Permeability to water of the roots of five herbaceous species. New Phytol. 72:547–555.

Scholander, P.F., H.T. Hammel, E. Hemmingsen and W. Carey. 1962. Salt balance in mangroves. Plant Physiol. 37:722–729.

Scholander, P.F., E.D. Bradstreet, H.T. Hammel and E.A. Hemmengsen. 1966. Sap concentrations in halophytes and some other plants. Plant. Physiol. 41:529–532.

Scholander, P.F. 1968. How mangroves desalinate seawater. Physiol. Plant. 21:258–268.

Shalhevet, J., E.U. Mass, G.J. Hoffman and G. Ogata. 1976. Salinity and the hydraulic conductance of roots. Physiol. Plant. 38:224–232.

Thau, G., E. Block and O. Kedem. 1966. Water transport in porous and non-porous membranes. Desalination 1:129–138.

CHAPTER 7

Observations on water salinity in mangrove associations at two localities in Papua New Guinea

J.S. WOMERSLEY*

82, Richmond Road, Westbourne Park, South Australia, 5041

This short paper reports the results of observations made with a salinity meter in waterways of the Labu Lakes near Lae, Morobe Province and in the delta of the Purari River on the south coast of Papua New Guinea.

The Labu lakes are a body of water on the southern side of the mouth of the Markham River. The lakes have a narrow opening to the Huon Gulf. On the seaward side the lakes are bounded by a sand spit carrying the usual coastal vegetation of *Pandanus, Barringtonia asiatica, Intsia bijuga, Derris indica, Pterocarpus indicus, Spondias dulcis, Terminalia catappa* and other trees. The lakes are open water margined by a dense mangrove dominated by Rhizophora and Bruguiera. The inland side of the lakes consists of a swampy coastal plain several miles in width which is terminated by the sharp rise of the Herzog Mountains from which a number of small streams originate to flow into the lakes through the swampy plain. The lower portion of the swampy plain is subject to daily flooding associated with the tidal rise and fall in the lakes. The vegetation is typically that of the middle and inner zones of the mangrove. *Xylocarpus granatum* is locally common. Above the influence of the tidal effects the swamp water is slow moving being entirely run-off from the mountains. The vegetation here is a mixture of *Metroxylon rumphii* (sago) swamp and a swamp rain forest in which *Terminalia complanata*, various species of *Syzygium* and

other trees are plentiful.

The Labu lakes themselves were used during the 1939–1945 Pacific War as a safe harbour for barges and a small floating dock. Observations near the mouth of the lakes indicated that at the optimum time of the rise and ebb of the tide there was a strong surface flow. At other times during the rise and fall the surface water was rather calm. The Markham river spews a considerable volume of fresh water into the Huon Gulf. This visibly silt-laden water floats for some distance out to sea. As the tide is rising some silt laden water originating from the Markham enters the lakes. At all times the surface water in the lakes is of low salinity and marginally potable.

To investigate the nature of the waterflow in the Labu lakes a series of water salinity observations were made at 6 sites with repition of readings at low tide and at full tide. These sites, and the results obtained are described in Table 1. During the course of these observations it was visually apparent that the silt laden water which entered the main channel of the lakes did not penetrate to the edges of the lakes and to the mangrove vegetation. This surface water remained clear. The observations indicate that only at site 1/8 opposite the mouth of the lakes and site 2/7, 7A which is adjacent is there any appreciable salinity in the surface water. Generally at these sites the water from the surface to about 50 cms in depth is of low salinity and only

---

* Formerly Assistant Director, Botany, Department of Forests, Papua New Guinea 1946–1975.

H.J. Teas (ed), Physiology and management of mangroves.
© 1984 Dr W. Junk Publishers, The Hague. ISBN 90 6193 949 6. Printed in the Netherlands.

Table 1. Salinity in parts per 1000 NaCl Labu Lakes, near Lae, New Guinea North Coast.

Depth	Site 1	2	3	4	5	6	7	7a	8	9	10	11
Surface	11.7	4.5	2.6	2.1	1.2	3.8	6.5	11.4	8.7	4.4	8.1	8.3
0.5 m	16.6	17.2	13.0	11.7	1.3	14.2	13.0	15.4	14.0	13.0	12.3	10.6
1.0 m	17.4	17.8	19.4		18.5	19.3		20.5	18.4		16.8	11.1
1.5 m			25.9		23.0	23.6		23.6			17.0	20.8
2.0 m			28.5		24.5	27.1		29.2				210
2.5 m												
3.0 m												
Bottom	17.4	17.8		11.7					18.4	13.0	17.0	

Table 1. Site details for Labu Lakes.

1   South east bank near Labu Village, tide 3/4 full. *Rhizophora apiculata*, pure community up to 80′ high, with a few *Bruguiera gymnorrhiza*.

2   As for site 1 but east bank, readings 3 m from the bank, tide 3/4 full. *Rhizophora apiculata* with occasional *Xylocarpus granatum*, *Heritiera littoralis* and *Avicennia marina*.

3   Upstream on right hand branch of Labu Lakes at second main fork. Channel 20 m wide, sample 3 m from bank; tide rising to full. *Rhizophora apiculata* with occasional *Xylocarpus granatum*.

4   As for site 3 but further upstream, sample 2 m out from tree roots, tide rising to full. *Rhizophora apiculata* with occasional *Xylocarpus granatum* and *Aegiceras corniculatum*.

5   Limit of boat navigability in creek, much fallen debris. Creek 20 m wide, sample from mid point; tide approaching full. *Rhizophora apiculata*, *Xylocarpus granatum*, *Aegiceras corniculatum*, but *Brownlowia argentea* and *Diospyros sp.* just appeared.

6   West branch of lakes above mouth, creek 30 m wide, sampling 3 m from edge of mangrove; tide almost full. *Rhizophora apiculata* with infrequent *Xylocarpus granatum* and *Bruguiera gymnorrhiza*.

7   As for site 2, tide full.

7   As for site 2 but 5 m from edge of mangroves, tide full.

8   As for site 1, tide full.

9   Labu Lakes, main south east branch near *Sonneratia* point. Mud island with centre of *Rhizophora apiculata* up to 4 m high above high water level surrounded by *Scirpus littoralis*, tide full.

10   As for 9 but 12 m from *Sonneratia caseolaris* tree stump, tide full. *Rhizophora apiculata* overwhelming *Sonneratia caseolaris* previously growing on this site.

11   Mouth of main channel to the lakes; tide almost full but still flowing well. *Rhizophora* on the banks. South eastern side with much sand and non-mangrove vegetation.

below 1 m is there a sharp rise in salinity indicating an inflowing of higher density ocean water from the Huon Gulf. The observations indicate that the mangroves of the Labu Lakes exist in a lake of low salinity water which floats on the rising and falling flow of saline water brought in and withdrawn by the tides.

These observations were extended to the Purari estuary to examine the system operating where there is a strong flow of fresh water at all times even though there is a tidal rise and fall of about 1 m. 6 sites were selected covering the main stream and some connecting branches of the Purari distributors. The site details are given in conjunction with Table 2.

The picture is rather different here where there is a strong and constant outrush of river water indicating that the mangrove vegetation is not being subject to inundation by strongly saline water, even in the lower reaches of the Wapo river where the water salinity is about 55% of that of ocean water off the south coast of Papua New Guinea. If there is any inflowing of high salinity water then this occurs below 5 m depth. The fact that the surface water throughout the Purari mangrove, perhaps the greatest single extent of Rhizophoraceous mangrove anywhere in the world – 225,000 hectares of tidal forests, is suitable for drinking has long been known to the Papuan people and foresters, immortalised, at least within the records of the Forestry

Table 2. Salinity parts per 1000 NaCl Purari Delta, Papua South Coast

Depth	Site 12	13A	13B	14	15	16
Surface	4.5	19.2	18.8	4.7	4.2	0.1
0.5 m	4.5	19.3	18.8	4.7	4.2	0.1
1.0 m	4.5	19.4	19.0	5.0	4.9	0.1
1.5 m	4.6		19.2	4.7	5.2	0.1
2.0 m	4.7		19.3	5.7	5.6	0.1
2.5 m	4.7		19.3	6.8	6.2	0.1
3.0 m			19.5	7.1	6.5	
Bottom	4.7 (5 m)	19.4	19.5	7.1	7.1 (3.2 m)	0.1

No bottom 8 m

Table 2. Site details for the Purari delta.

12	Baimuru, adjacent to wharf. River bank with *Rhizophora apiculata* (small) *Sonneratia* and *Excoecaria agallocha*; tide near full.
13A	East branch of the Wapo River, mangrove forest dominated by *Rhizophora apiculata, Bruguiera gymnorrhiza, Xylocarpus granatum*; tide near full.
13B	Thirty metres from the bank at site 13A; tide near full.
14	Mina River, upper limit of *Nypa* and *Sonneratia* lining river banks; tide full.
15	Same as 14; falling tide.
16	Junction of Tiviri River and Pai-A River. Rainforest on banks, river tidal; tide full, just turning.

Department by M.F.C. Jackson with the remark, that, at the end of a loathsome day surveying in the mangrove one could go from deep depression to mild optimism with a couple of tots of rum and Purari water.

Similar, but much cruder observations were made in January 1979 in the Sunderbans area of Bangladesh where the river system consists of the estuarine distributaries of the Ganges, Meghna and Brahmaputra. Using what can be described as the 'taste test' the surface water at both low and high tide was not unpleasantly saline. It is well known that within the Sunderbans deer, tiger and monkey drink from the tidal creeks. The conclusion is that at least in the two sites observed in Papua New Guinea and in the Sunderbans the water which inundates the pneumatophores and lower stilt roots of the mangrove trees is only slightly saline. While fresh water is not essential for the growth of the species of mangrove trees, for example in Fairfax Harbour Port Moresby, the great areas of luxuriant mangrove exist in a situation where a lake of low salinity water is floating on a tidal flow and ebb of high salinity water or the outrush of fresh water in the river systems is 'dammed' by the rising tide and released by the ebbing tide.

CHAPTER 8

The effects of 2,4-D on the growth of *Rhizophora stylosa* Griff. seedlings

P. CULIC

Biology Department, Capricornia Institute of Advanced Education, Rockhampton Qld., Australia

Abstract. A laboratory study was conducted to determine the effects of 2,4-D amine salt on the growth of seedlings of the mangrove, *Rhizophora stylosa* Griff. Seedlings were allowed to grow for four weeks (up to the emergence of the first pair of leaves), and leaf and soil treatments of 2,4-D were applied (leaf: 0.0125–2.50 kg/ha; soil: 0.3125–62.50 kg/ha).

Sensitivity of seedlings to the herbicide was evidenced by an increase in first internode length, retardation of dry matter accumulation and changes in chlorophyll synthesis.

Introduction

Mangrove plant communities have attracted much scientific attention through their characteristic adaptions to relatively harsh environmental conditions (Lugo & Snedaker 1974). These woody perennials are adapted to a particular environment in which few higher plants can compete and occupy considerable areas of the world's tropical and subtropical coastlines and estuaries. Mangroves have both economic and commercial importance (Navalkar 1961; Macnae 1968; Rollet 1975; Kader 1979), and they are of environmental value as buffers against erosion and in their role as land-builders (Bird 1971; Carlton 1974).

Oil spills, land reclamation, organic waste pollution and dumping of rubbish are potential threats to mangrove communities. Other threats include organic chemical pollutants such as pesticides and herbicides. Most pesticides are biologically active chemicals that are toxic to one or more forms of life (von Rumker *et al.* 1975); herbicides are also biologically active chemicals and their use has caused much controversy. Reliable figures on world pro-

duction and use of herbicides are hard to obtain, but in the United States of America alone in 1972 over 200 thousand tons of herbicides were produced (von Rumker *et al.* 1975). The discovery of 2,4-D (2,4-dichlorophenoxyacetic acid) during World War II initiated a revolution in chemical weed control (Peterson 1967). In Australia, the herbicide 2,4-D was first introduced in the late 1940's and its use at present is in the order of 2,000–2,500 tons per year (L. Jones 1978, Personal Comm.)

Serious doubts have been raised about the safety of 2,4-D with respect to contamination of native and commercial vegetation. Scientific studies on the use of 2,4-D have shown various adverse effects on plants of economic importance (Wilde 1951; Johanson & Muzik 1961; Waddington *et al.* 1976). The military use of 2,4-D and other herbicides in South Vietnam since 1962 has caused much concern among biologists (Tschirley 1969; Orians & Pfeiffer 1970; Westing 1971; Aaronson 1971; Boffey 1971). Susceptibility of mangroves to 2,4-D was first investigated by Ivens (1957). First signs of herbicide effect were noted approximately 3 weeks after ap-

H.J. Teas (ed), Physiology and management of mangroves.
© *1984 Dr W. Junk Publishers, The Hague. ISBN 90 6193 949 6. Printed in the Netherlands.*

plication and by seven months extensive defoliation had occurred. Studies by Truman (1961), Ross (1975) and Teas & Kelly (1975) support these findings.

Hazard to mangroves from 2,4-D treatment of agricultural hinterlands might occur from either spray drift or from run-off water that reaches estuaries. 2,4-D has been detected in streams after spraying, (Norris 1969; Choi Lee *et al.* 1976) and after direct application to streams to control water weeds (Frank 1972). The concentration of herbicide which might reach mangroves would be determined by several factors including: herbicide dosage, area treated, soil binding and elution, rainfall, tidal and other dilution (Teas 1976).

Rhizophora stylosa can be considered as one of the most successful tidal representatives in Eastern Australia. The tree is viviparous and produces seedlings consisting of an elongated hypocotyl with a plumule. Leaf development does not occur until roots become established in the soil.

The purpose of this study is to describe the effects of 2,4-D amine salt (50% commercial formulation) on the growth of seedlings of *Rhizophora stylosa,* Griff.

Materials and methods

All plant material was collected in early December from fruiting *R. stylosa* trees located 6 km upstream from the mouth of the Fitzroy River (Rockhampton). Only those seedlings where the hypocotyl length was 25–35 cm were collected. Mud was collected from the same site and was placed into germination troughs (80 cm × 20 cm × 20 cm). Each trough was partitioned, using a polystyrene sheet giving 2 plots; each 40 cm long. Ten (10) seedlings were planted, in each plot, in two rows of 5 approximately 3 hours after collection. The seedlings were inundated daily with half strength sea water.

Application of herbicide

Seedlings were allowed to grow for 4 weeks to allow the emergence of the first set of leaves and the establishment of a root system.

Aqueous solutions of 2,4-D were made to the following concentrations: 5 ppm; 50 ppm; 500 ppm; 1000 ppm. 2,4-D treatments were given to the seedlings by either leaf or soil application. Leaf application to the adaxial surface was by means of a spray atomiser, with a plastic shield placed under the leaves to prevent contamination of the soil. Two (2) sprays from an atomiser with the equivalent of 1.0 ml of herbicide solution were sufficient to completely wet the surface of one leaf. Soil application was by adding 500 ml of herbicide solution to each plot. Untreated seedlings were used as controls. 2,4-D dosages used were calculated using the following formula.

$$\text{2,4-D Dosage} \atop \text{(kg/ha)} = \frac{V \cdot \{ppm\}}{10^5 \cdot A}$$

where

V	= total volume of 2,4-D solution applied to seedlings.
{ppm}	= concentration of 2,4-D in parts per million.
A	= surface area of plot (m²).

Observations and measurements

Ten (10) seedlings were sampled from untreated and treated plots before the application of 2,4-D and then each month, for 3 months after application. Measurements were taken of the first internode length (Fig. 1). Additionally leaves from each seedling were weighed to determine the fresh weight and then oven dried at 60° C for 72 hours to determine the dry weight (O.D.W.).

Leaf surface area was calculated by incorporating length and width values into the following equation:

$$\log(S.A.) = 0.8468 + (0.0051.L) + (1.3188.\log B)$$

where:
S.A. = Surface Area (cm²),
L = Length (cm),
B = Breadth (cm).

The equation was derived by establishing an em-

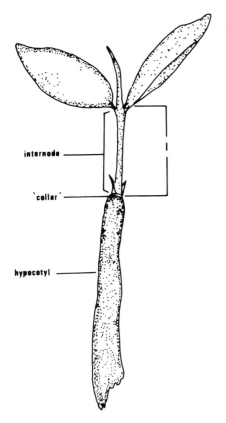

internode

'collar'

hypocotyl

Fig. 1. Diagram showing length of 1st internode from the 'collar'.

pirical relationship using multiple regression analysis between leaf length, breadth and surface area for forty (40) leaves. (A multiple correlation coefficient of 0.97 was obtained). Using a paired t test, measured and empirically determined surface area of leaves were compared. This was found to be non-significant.

Similar plots and treatments were used to determine changes in total chlorophyll in leaves.

From each plot, 5 seedlings were randomly selected and used for chlorophyll analysis. Chlorophyll analysis entailed the collection of leaves which were first weighed and then ground in Acetone (Anala®) using a precooled mortar and pestle. The extract was then decanted, filtered into a 50 ml volumetric flask, and made to volume. This extract was centrifuged using a Beckman Model J-21 Centrifuge (set for 20° C) for 10 minutes at 700 r.p.m. The chlorophyll levels in the supernatant were measured using a Perkin-Elmer 124, Double Beam

Spectrophotometer. The chlorophyll concentrations were determined at the wave-lengths of 647 nm and 644 nm and using the equation of Jeffrey & Humphrey (1975).

Results

In the interpretation of data only differences falling outside the 95% confidence interval were considered. These 95% c.i. values were determined using the Root Mean Square (R.M.S.) of standard error for the data points concerned as the estimate for s, the error bounds are determined by the standard technique recorded in Sokal and Rohlf (1969 p. 146) where

$$\bar{y} + t_{0.05} (136) \ \frac{s}{\sqrt{10}}$$

Figures 2(a) and (b) show changes in the first internode length with time for different treatments of 2,4-D. At 30 and 60 days leaf treated seedlings showed a increase in the internode length for all treatments relative to control. This increase was still evident at 90 days but only for the two lowest dosages. Seedlings subjected to soil application showed a similar increase but only for the three (3) highest dosages at 30 days, and at 60 and 90 days only for second lowest dosage. Internode lengths were not recorded after 60 days for seedlings subjected to the two highest dosages of 2,4-D, for both applications. The internodes of these seedlings exhibited marked wrinkling and shrinkage, this indicating cell degeneration.

Figures 3(a) and (b) show changes in leaf dry matter accumulation with time. Leaf treated seedlings at 30 days showed a decline in dry matter accumulation for the two highest dosages but by 90 days this decline was evident for all treatments. A similar trend occurred with soil treated seedlings except that by 90 days, seedlings treated with the highest dosage of 2,4-D were defoliated.

Figures 4(a) and (b) show changes in leaf biomass per unit area. Leaf treated seedlings showed an apparent increase in biomass per unit area, for the three highest dosages of 2,4-D at 30 days, but

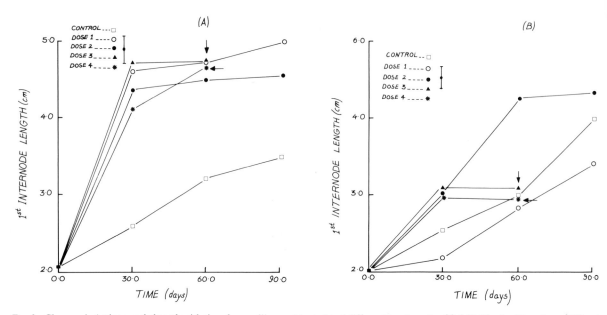

Fig. 2. Changes in 1st internode length with time for seedlings subjected to 4 different treatments of 2,4-D. Vertical line shows 95% c.i. for data points. (↑ : Internode cell degeneration following this point).

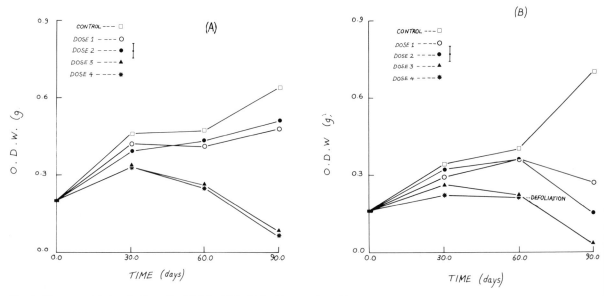

Fig. 3. Changes with time in Oven Dry Weight (O.D.W.) of leaves of seedlings subjected to 4 different treatments of 2,4-D. Vertical line shows 95% c.i. for data points. (A) Leaf Application (kg/ha): Dose 1 = 0.0125; Dose 2 = 0.125; Dose 3 = 1.25; Dose 4 = 2.5. (B) Soil Application (kg/ha): Dose 1 = 0.3125; Dose 2 = 3.125; Dose 3 = 31.25; Dose 4 = 62.5.

by 90 days all treated seedlings showed a decline. With soil treatments, all seedlings showed a relative decline in leaf biomass per unit area at 60 days with defoliation occurring with the highest dosage by 90 days.

Figures 5(a) and (b) show changes in total chlorophyll content in leaves. In the leaf treated seedlings, those given the two intermediate dosages of 2,4-D had higher chlorophyll values at 30 days. At 90 days seedlings treated with the highest

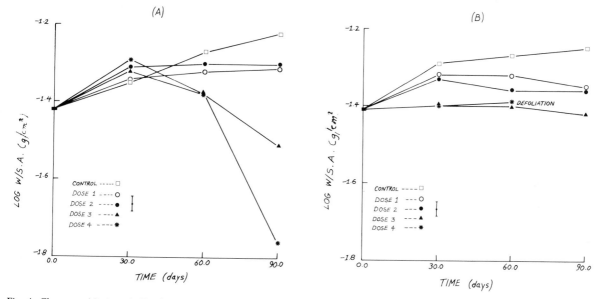

Fig. 4. Changes with time, in Fresh Weight/Surface Area (g/cm²) of leaves of seedlings subjected to 4 different treatments of 2,4-D. Vertical line shows 95% c.i. for data points.

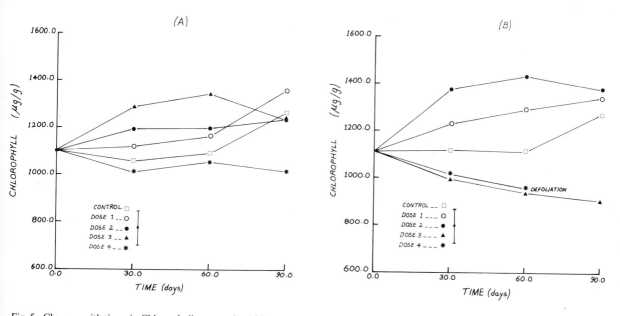

Fig. 5. Changes with time, in Chlorophyll content (μg/g) in leaves of seedlings subjected to 4 different treatments of 2,4-D. Vertical line shows 95% c.i. for data points. (A) Leaf Application (kg/ha): Dose 1 = 0.0125; Dose 2 = 0.125; Dose 3 = 1.25; Dose 4 = 2.5. (B) Soil Application (kg/ha): Dose 1 = 0.3125; Dose 2 = 3.125; Dose 3 = 31.25; Dose 4 = 62.5

dosage of 2,4-D showed a decline in chlorophyll value relative to all other treatments. Soil treated seedlings were found to have higher chlorophyll values at the lower dosages at 30 days, while at the higher dosages of 2,4-D there was a decline in chlorophyll values relative to control. Seedlings treated with the highest dosage were defoliated by 90 days.

It is evident from these results that *R. stylosa* seedlings are susceptible to the herbicide 2,4-D.

This susceptibility is obvious in all parameters studied but more so using the first internode measurement; this parameter showed relatively more significant changes 30, 60 and 90 days after treatment than those observable in the remaining three parameters.

A comparison between leaf and soil application indicates a greater degree of susceptibility of seedlings to soil treatments of 2,4-D. This is best observed with changes in O.D.W. and chlorophyll values in the leaves and furthermore only soil treated seedlings were defoliated with the highest dosage of 2,4-D.

Discussion

The mangrove seedlings in control, although maintained in artifical conditions appeared to grow normally. Patil (1964) successfully cultivated 8 species of mangrove seedlings in laboratory-type conditions and believed that further growth beyond the early seedlings stages was possible.

Published experimental tolerance levels of *R. mangle* L. to herbicides range between 4.4 kg/ha of 2,4-D and 1.6 kg/ha picloram, all seedlings being killed after 40 days (Walsh 1973) to a lethal dose of 13 kg/ha (Teas 1976). 'No permanent effect' dosages range between 0.44 kg/ha of 2,4-D and 1.16 kg/ha picloram (Walsh 1973) to <0.01 kg/ha (Teas 1976). Due to experimental differences such as, combined herbicide treatment, method of treatment and a different species of mangrove used by the above authors it is difficult to make a comparison with their work; however death was only noted at 62.50 kg/ha for soil treated seedlings after 60 days while seedlings treated with dosages of 2,4-D as low as 0.0125 kg/ha applied to the leaves or 0.3125 kg/ha applied to the soil exhibited marked retardation in growth after 60 days.

This work demonstrated 2,4-D's marked effects on the growth of *R. stylosa;* such results may come from direct interference with the division, enlargement or differentiation of growing cells (Ashton & Crafts 1973). Increased internode length in seedlings treated supports this claim and suggests 2,4-D interference with normal cell enlargement and division (Kaufman 1955; Key *et al.* 1966; Audus 1976).

Negative responses relative to control, to dry matter accumulation and changes in biomas per unit area in leaves of treated seedlings are in accordance with findings of seedling growth retardation and indicate cell degeneration with high dosage of 2,4-D. Relatively lower dosages caused a retardation in normal leaf growth but no permanent effects were apparent. Hallam (1970) working with leaves of *Phaseolus vulgaris* L. var Canadian Wonder found that 2,4-D caused a breakdown in the membranes of cells of the epidermis, palisade and mesophyll. Similarly Walsh (1973) found that seedlings of *R. mangle* treated with 4.4 kg/ha of active ingredient 2,4-D showed histological abnormalities in the leaves where the cell wall continuity began to break in the hypodermis, palisade and spongy parenchyma.

Negative responses relative to control to chlorophyll production in leaves of seedlings treated with high dosages of 2,4-D are in accordance with findings by Hallam (1970) and White & Hemphill (1972) and with the hypothesis of cell degeneration. However the initial stimulation of chlorophyll production by 2,4-D found here is difficult to reconcile with the above hypothesis. Reduction of size of vascular bundles in leaves (Eames 1949) increase in the depth of palisade tissue (Bradley *et al.* 1968) and increase in cell and nucleus size (Audus 1976) may be responsible for this initial stimulation of chlorophyll production.

My experiments indicate that relatively low concentrations of 2,4-D affect normal growth of *R. stylosa* seedlings. 2,4-D contamination of *R. stylosa* communities may significantly retard seedling development and if residues persist in the soil new seedling establishment and colonization could be inhibited.

As far as I am aware few workers have reported values for herbicide concentrations in run-off from herbicide treated lands (Bovey *et al.* 1974; Choi Lee *et al.* 1976). While the concentrations reported are certainly lower than those used in my present work, care should be exercised to ensure there is no hazard to mangroves, especially where 2,4-D application is near estuaries, until more work has been done on 2,4-D run-off from agricultural districts.

Acknowledgements

The author extends his sincere thanks to Dr. J. Kowarsky and Dr. A.P. Mackey for their helpful criticism in the preparation of this manuscript. Thanks also go to the Department of Biology technical staff for their support and to the Capricornia Institute of Advanced Education for their financial assistance.

Literature cited

Aaronson, T. 1971. A tour of Vietnam. Environment 13:23–43.

Ashton, F.M. and A.S. Crafts, 1973. Mode of action of herbicides. Wiley-Interscience, New York, USA.

Audus, J.L. 1976. Herbicides, I. Academic Press, New York, USA.

Bird, E.C.F. 1971. Mangroves as land builders. Victorian Naturalist 88:189–197.

Boffey, P.M. 1971. Herbicides in Vietnam: AAS study finds widespread devastation. Science 171:43–47.

Bovey, R.W., E. Burnett and C. Richardson. 1974. Occurrence of 2,4,5-T and picloram in surface runoff water in the blacklands of Texas. Journal of Environmental Quality 3:61–64.

Bradley, M.V., J.C. Crane and N. Marei. 1968. Some histological effects of 2,4,5-trichlorophenoxyacetic acid applied to mature apricot leaves. Botanical Gazette 129:231–238.

Carlton, J.M. 1974. Land building and stabilization by mangroves. Environmental Conservation 1:285–294.

Choi, Lee. K., S.S. Que Hee and R.G. Sutherland. 1976. 2,4-D levels in the south Saskatchewan river in 1973 as determined by a glc method. Journal of Environmental Science and Health B11:175–183.

Eames, A.J. 1949b. Comparative effects of spray treatments with growth-regulating substances on the nut grass, *Cyperus rotundus* L., and anatomical modifications following treatment with butyl 2,4-dichlorophenoxyacetate. American Journal of Botany 36:571–584.

Frank, A.P. 1972. Herbicidal residues in aquatic environments. Pages 135–148. In: Fate of organic pesticides in the aquatic environment R. Gould, ed. Advances in Chemistry Series III.

Hallam, N.D. 1970. The effect of 2,4-dichlorophenoxyacetic acid and related compounds on the fine structure of the primary leaves of *Phaseolus valgaris*. Journal of Experimental Botany 21:1031–38.

Ivens, G.W. 1957. Arboricides for killing mangroves. West African Rice Research Station Periodical Science Report, 8.

Jeffrey, S.W. and G.F. Humphrey. 1975. New spectro photometric equations for determining chlorophylls a, b, c, and c_2 in higher plants, algae and natural phytoplankton. Biochemie and Physiologie der Pflanzen 167:191–194.

Johanson, N.G. and T.J. Muzik. 1961. Some effects of 2,4-D on wheat yield and root growth. Botanical Gazette 122:188–194.

Kader, R.A. 1979. How Malaysia manages its mangroves. Forest and Timber 15:7–8.

Kaufman, P.B. 1955. Histological responses of the rice plant (*Oryza sativa*) to 2,4-D. American Journal of Botany 42:649–659.

Key, J.L., C.Y. Lin, E.M. Gifford Jr. and R. Dengler. 1966. Relation of 2,4-D-induced growth abberrations to changes in nucleic acid metabolism in soybean seedlings. Botanical Gazette 127:87–94.

Lugo, A.E. and S.C. Snedaker. 1974. The ecology of mangroves. Annual Review of Ecology and Systematics 5:39–64.

Macnae, W. 1968. Mangroves and their fauna. Australian Natural History 16:17–21.

Navalkar, B.S. 1961. Importance of mangroves. Tropical Ecology 2:89–93.

Norris, L.A. 1969. Herbicide runoff from forest lands sprayed in the summer. Research Progress Report, Western Society of Weed Science 24–26 p.

Orians, G.H. and E.W. Pfeiffer. 1970. Ecological effects of the war in Vietnam. Science 168:544–554.

Patil, R.P. 1964. Cultivation of mangroves seedlings in pots at Allahabad, U.P. Science and Culture 30:43–44.

Peterson, G.E. 1967. Discovery and development of 2,4-D. Agricultural History 41:243–253.

Rollet, B. 1975. Utilization of mangroves. Journal D'Agriculture Tropicate et de Botanique Appliquee 22:203–235.

Ross, P. 1975. The mangrove of south Vietnam: The impact of military use of herbicides. Proceedings of the International Symposium on Biology and Management of Mangroves 2:695–709.

Sokal, R.R. and F.J. Rohlf. 1969. Introduction to biostatistics. W.H. Freeman and Company, San Francisco.

Teas, H.J. 1976. Herbicide toxicity in mangroves. Report, US Environmental Protection Agency Office of Research and Development, EPA-600/3-76-004, 1–33 p.

Teas, H.J. and J. Kelly. 1975. Effects of herbicides on mangroves of south Vietnam and Florida. Proceedings of the International Symposium on Biology and Management of Mangroves 2:719–727.

Truman, R. 1961. The eradication of mangroves. Australian Journal of Science 24:198–199.

Tschirley, F.H. 1969. Defoliation in Vietnam. Science 163:779–786.

von Rumker, R., E.W. Lawless and A.F. Meiners. 1975. Production, distribution, use and environmental impact potential of selected pesticides. Report for, Council on Environmental Quality, Washington, DC.

Waddington, J., J. Gebhardt and D.A. Pulkinen. 1976. Forage yield and quality alfalfa following late fall applications of 2,4-D or 2,4-DB. Canadian Journal of Plant Sciences 56:929–943.

Walsh, G.E. 1973. Effects of herbicides on seedlings of the red mangrove, *Rhizophora mangle* L. Bioscience 23:361–364.

Westing, A.H. 1971. Ecological effects of military destruction on the forests of south Vietnam. Bioscience 21:893–898.

White, J.A. and D.D. Hemphill. 1972. An ultra-structural study of the effects of 2,4-D on tabacco leaves. Weed Science 20:478–481.

Wilde, M.H. 1951. Anatomical modifications of bean roots following treatment with 2,4-D. American Journal of Botany 38:79–91.

CHAPTER 9

A realistic approach to the use and management of mangrove areas in Southeast Asia

ARMANDO A. de la CRUZ

Department of Biological Sciences, Mississippi State University, Mississippi State, MS 39762, USA

Abstract. In Southeast Asia and other developing regions of the tropical world, the coastal zone will inevitably be used and exploited for food production and habitation. To attain these goals, and at the same time preserve the integrity of the mangrove areas, requires a management strategy that is unique to the Asian situation. Coastal mangrove areas should be classified on the basis of their present state of alterations into *Reserved Mangroves* which should be protected for their natural history value; *Managed Mangroves* which should be developed for optimum food, fiber and fish production; and *Altered Mangroves* which should be restored by reforestation and/or converted to productive land uses. The basis for management is a comprehensive understanding of the ecological structure and function of the mangrove ecosystem.

Introduction

Food production is the number one national goal of every developing nation to feed its rapidly growing population. This means putting into productive use every available space that has the potential to produce food. In Southeast Asia and elsewhere in the world, the coastal zone inevitably will be used and exploited to grow grain and fiber, and to produce protein. A major environment threatened by the rapidly ensuing utilization of the coastal zone are the mangrove tidal forests and nipa palm swamps. The ecological values of these ecosystems revolve about their naturally high energy production capacity (Christensen 1977), their contributions to estuarine and marine fertility (Heald and Odum 1970; Macnae 1974), and their various ecosystem functions particularly as nursery habitat to numerous fish and shellfish (de la Cruz 1976; 1979). To achieve the goal of food production in coastal regions and at the same time preserve the ecological integrity of the mangrove areas, will require a man-

agement scheme that is realistic and workable within the conditions that now exist in Southeast Asia. The aim of this paper is to review the present utilization of mangrove resources and to suggest a method for management.

Status of mangrove resources in ASEAN countries

Estimates of the areal extent of mangrove resources in Indonesia, Malaysia, Philippines, and Thailand are available (Table 1) but the exact hectarage is not accurately known. A systematic survey of mangrove areas is urgently needed in each country. Precise and rapid measurement is feasible with the use of aerial photographs and satellite imagery (Umali 1977).

Informations from available literature and field-work conducted in 1977 and 1978 revealed that generally, the mangrove resources in Southeast Asia can be categorized as follows:

1 Natural stands of primary and secondary for-

H.J. Teas (ed), Physiology and management of mangroves.
© *1984 Dr W. Junk Publishers, The Hague. ISBN 90 6193 949 6. Printed in the Netherlands.*

Table 1. Approximate hectarage of mangrove resources in ASEAN countries.

Country	Hectares of Mangrove Forest
Indonesia	1,000,000[1]
Malaysia	655,572[2]
Philippines	251,577[3]
Thailand	307,715[4]

[1] Estimate values communicated by Dr. Aprilani Soegiarto of the Indonesian Institute of Oceanology and Dr. Kuswata Kartawinata of Herbarium Bogoriense.
[2] Calculated from figures reported in the Proceedings of the Workshop on Mangrove and Estuarine Vegetation held at Universiti Pertanian Malaysia Fakulti Perhutanan, Serdang, Selangor in December 10, 1977.
[3] One of the values reported from a range of 100,000 – 400,000 ha during the National Symposium and Workshop on Mangrove Research and Development held in Manila, July 28–30, 1977.
[4] Average of six values obtained from Dr. Sanit Aksornkoae of Kasetsart University and from the literature.

ests which are old but healthy and reasonably undisturbed;

2 Young stands of mainly secondary forests which are vigorously growing and subject to periodic harvest or more frequent pruning;

3 Mangrove areas which have been completely altered and are either idly eroding or converted to other land uses.

The areal extent of mangroves that fall under any one of these categories is difficult to obtain and probably not known. In Indonesia, there still remains extensive stands of primary forest in Celebes, Kalimantan and Sumatra: but the mangroves on the island of Java has been permanently altered and converted to 'tambak' areas. Isolated tambaks are reclaimed; for example in Muara Blanakan, Sukamundi, some 800 ha of tambak are managed by the Indonesian Forest Planning Service as *Rhizophora/Avicennia* mangrove plantations. Probably, Malaysia has the highest hectarage of mangrove forest under silvicultural development. The Malaysian Forest Department has a well documented scheme of mangrove forest management that dates back to the early 1900's (Watson 1928). Some 40,874 ha of the 173,614 ha of mangroves in Sarawak has been declared forest reserves and protected forests (Chai 1980). In Sabah, heavy clearing of mangroves for export as wood chips took place in much of 365,345 ha of forests in the 1970's (Chim 1980). A major development in mangrove resources utilization in the Philippines is conversion to fish ponds. Umali (1977) estimated that about 24,200 ha were denuded between 1967–1977. Datingaling (1977) reported that 90,571 ha were at that time under government leases primarily for fish pond operation. It may be true that more than one half of the mangroves in the Philippines are under some kind of alteration. The mangrove resources in Thailand is also under heavy pressures. Of the 307,715 ha estimated area of mangrove, 167,000–179,000 ha are leased to lumber concessions and 40,880 ha have been cleared for shrimp farming. A Mangrove Forest Reserve has been designated in Kor Ta Lu Tao in southern Thailand which represents an attempt by the Thai government to preserve some of its mangrove areas.

It is inevitable that coastal areas with mangrove vegetation will be developed for food production and habitation purposes in Southeast Asia. It is also clear that large areas have already been permanently altered, that some areas are managed for food and fiber production, and alarmingly small and patchy hectarage may still be in their pristine condition. What is, therefore, needed is a management strategy that will address all the three mangrove categories described above.

Strategy for management

An obvious requirement for a management scheme adapted to the Asian situation is an accurate survey map of the mangrove resources of each country. From aerial photos supplemented with satellite imagery, it would be a simple process to distinguish permanently altered coastal areas from healthy mangrove stands. Ground truth information will delineate primary mangrove forest from secondary growth, and present forestry statistics will determine mangrove stands under silviculture operations.

From numerous discussions with workers in the field and from various unpublished references, it appears that mangrove resources can be categorized as follows:

1 Reserved Mangrove;
2 Managed Mangrove;
3 Altered Mangrove.

Reserved mangrove

These are healthy stands of primary and/or secondary forests that should be protected through properly legislated national policies. Several of these reserved forests should be declared Mangrove Reserved all over the country. They are to be effectively protected from any form of alteration. Accessible areas can be cautiously used for scientific and educational purposes. The government can decide how many Mangrove Reserves and how large areas are to be protected and their location to be determined for their ecological function relative to adjacent ecosystems.

Managed mangroves

These are mangrove forests under silvicultural management for wood products production. These forests should be managed in the same manner terrestrial forests are managed. This will involve a systematic scheme of rotation felling, reforestation requirements, nursery studies to develop high yielding varieties of mangrove species, new mechanics of harvest which are both efficient and will inflict less damage to the waterlogged substrate. Foresters have to acquire the know-how of sound silvicultural practices applicable to coastal wetlands.

Altered mangroves

These are permanently altered mangrove areas that have been completely cleared and/or converted to other land uses. These areas can be developed into salinas, rice farms, fish and shrimp ponds, or rehabilitated back to mangrove forest plantations. The concept of multiple usage of this type of resource should be employed.

The Indonesians have a systematic reforestation program of the idle tambak areas of Java. For example, at Muara Blanakan in Sukamundi, West Java, the Indonesian Forest Planning Service and Forestry Department have established mangrove plantations revegetated with *Avicennia, Rhizophora, Ceriops, Sonneratia* and *Pluchia indica*. The 780-hectare tambak field is diked into hectare-size plots and planted as follows:

1 *Avicennia* is planted from seeds in January; 3–4 seeds per mound at 2–3 m spacing and harvested after 5 years.
2 *Rhizophora* is planted from fruit with 2–3 m spacing; harvested 20–30 years depending on preferred size.
3 *Ceriops* is planted from fruit with 2–3 m spacing; harvested after 4 years.
4 *Sonneratia* is planted from stems with 2–3 m spacing; harvested after 4 years.

The national plan of the reforestation programs for idle tambak areas follow the following general schemes:

1 Reforestation priority is given to the shoreline or areas most adjacent to tidal influx.
2 The hectare–size plots are diked and surrounded by regularly maintained 2–3 m wide canals which served as tidal drainage and natural fish ponds where both *Tilapia* and *Chanos* are reared from seed fish without supplementary feeding, fertilizer addition or pesticide application. Fish harvest is done every 4–5 months yielding 200–300 kg/ha for *Chanos* and 800–1000 kg/ha for *Tilapia*.
3 Plots already diked and channelized and awaiting planting are temporarily used as salt ponds during dry season yielding 250 kg/day/5 m² subplots; and flooded as fish ponds during the rainy season.
4 The canals are also used in some areas for rearing of the mangrove snake *Cerberus rhynchops* which is periodically collected, chopped and fed to ducks.

The Indonesians have what appear to be a working management program for the permanently damaged mangrove areas and an impressive multi-use scheme for the reforested tambak areas.

Preservation, conservation, and production intensification

The management approach suggested above implies three important environmental concepts. The establishment of Mangrove Reserves will insure the preservation of some forests that will continue to possess a natural history value and will become a permanent part of the country's natural heritage. The concept of preservation is non-use; conservation on the other hand, means wise-use. This implies that the silviculture of Managed Mangroves should aim at maximum yield in minimum time. Conservation also implies that the mangrove forests under cultivation must continue to exist as a renewable resource.

The management scheme being suggested here discourages clearing and opening of more mangrovelands, that is, extensification of alteration. There is already a large hectareage of the so-called tambak areas. The strategy is to decide which, where and how much areas are to be developed for saline agriculture, aquaculture, or silviculture and then intensify the production of these areas. At the present level of fish, shrimp or rice yields in existing tambaks, governments can easily double or triple food production capacities of the coastal zone without cutting down another mangrove tree.

Acknowledgements

This work was supported by a grant from the US National Science Foundation Division of International Programs (NSF/SEED No. INT7707133).

Literature cited

Chai, P.K. 1980. Mangrove Forests of Sarawak. In Workshop on Mangrove and Estuarine Vegetation, 10 December 1977. Srivastava, P.B.L. and R.A. Kader (eds.), pp. 1–5. Fakulti Perhutanan, Universiti Pertanian Malaysia, Serdang, Selangor, Malaysia.

Chim, L.T. 1980. Mangrove Forest of Sabah. In Workshop on Mangrove and Estuarine Vegetation, 10 December 1977. Srivastava, P.B.L. and R.A. Kader (eds.), pp. 6–38. Fakulti Perhutanan, Universiti Pertanian Malaysia, Serdang, Selangor, Malaysia.

Christensen, B. 1977. Primary production of mangrove forests. Proceedings of International Workshop on Mangrove and Estuarine Area Development for the Indo-Pacific Region. Publ. Philippine Council for Agriculture and Resources Research (PCARR), Los Baños, Laguna, Philippines. pp. 131–135.

Datingaling, B. 1977. Aquaculture potentials of the mangrove area. Proceedings National Symposium/Workshop on Mangrove Research and Development, Publ. Philippine Council for Agriculture and Resource Research (PCARR). Los Baños, Laguna, Philippines.

Cruz, A.A. de la, 1976. The functions of coastal wetlands. Association of Southeastern Biologists Bulletin, 23:179–185.

Cruz, A.A. de la, 1979. The functions of mangroves. In Mangrove and Estuarine Vegetation in Southeast Asia. Publ. BIOTROP Special Publ. No. 10, Bogor, Indonesia. pp. 125–138.

Heald, E.J. and W.E. Odum. 1970. The contribution of mangrove swamps to Florida fisheries. Proc. Gulf and Carrib. Fish. Inst., 22:130–135.

Macnae, W. 1974. Mangrove forests and fisheries. FAO Publ. IOFC/DEV/74/34., FAO, Rome. p. 7.

Teas, H.J. 1979. Silviculture with saline water. In The Biosaline Concept. Hollaender, A. (ed.), pp. 117–161. Plenum Publ. Corp.

Umali, R. 1977. Present status and distribution of mangrove areas in the Philippines. Proc. National Symposium/Workshop on Mangrove Research and Development, Publ. Philippine Council for Agriculture and Resource Research (PCARR). Los Baños, Laguna, Philippines.

Watson, J.G. 1928. Mangrove forests of the Malay Peninsula. Malayan Forest Record No. 6, Forest Department, Kuala Lumpur, Malaysia, 225 p.

CHAPTER 10

The mangrove ecosystem in Indonesia, its problems and management*

APRILANI SOEGIARTO

National Institute of Oceanology, Jakarta – Indonesia

Abstract. The mangrove forests cover 3.6 million hectares along the coastal areas of Indonesia. Ecologically they are very important. They serve to protect beaches from sea abrasion, as well as spawning, breeding and nursery grounds of many economically important species. Recently, however, this ecosystem has received heavy pressures as a result of increasing human activities. The mangrove trees are used as firewood, charcoal, chipboard, raw materials for paper and in a rather limited way are used for timbers. In the last few years, there are an increasing desire and efforts to convert mangrove ecosystem into various uses, such as for rice fields, fish ponds, settlements, ports and industrial estates.

Studies have been carried out by universities and research institutions in Indonesia in order to fully understand the ecological functioning of the mangrove ecosystem and to find ways and means to manage it. This paper summarizes the problems and efforts, both in research and legislation, that have been encountered in trying the manage the mangrove ecosystem in Indonesia.

Introduction

The rich and diversified life of the coastal areas has been an important source of food for the Indonesian people for centuries. Fishes, crustaceans, molluscs, and seaweed are a few examples of these resources. In addition, the minerals and hydrocarbon resources currently have been tapped from the shallower parts of our seas.

Aside from the renewable and non-renewable resources, our coastal waters have many other roles, such as for inter-island trade, transportation, communication, recreation and tourism. Lately, however, as undesirable side effects of our strong efforts in economic development, the coastal environment experiences severe pressures, either directly or indirectly, resulting in the general degradation of certain coastal areas, including the mangrove ecosystem.

The state of knowledge of mangrove in Indonesia

Mangrove forests grow in humid tropical coastal areas of the world. One of the centers for distribution is the South-east Asian region. Mangrove forests normally develop along the protected coastal areas with muddy to sandy bottoms. But in some cases they are found on wave swept rocky coasts. In Indonesia mangroves develop well along the inner facing coast lines of most of the large islands. The following is a brief review of our knowledge on the

* Paper prepared for the Second International Symposium on Biology and Management of Mangrove, Port Moresby-Papua New Guinea, July 20–26, 1980.

Indonesian mangrove ecosystem, in part taken from Soegiarto (1980).

Areal coverage

From earlier literatures the areal coverage of the mangrove forest in Indonesia was listed ranging from one to two million hectares. However, through intensive surveys and mapping using satellite imagery as well as aerial photography the latest figures were reported as 3.6 million hectares mangrove forest areas with about 60 percent (2.1 million hectares) as forest stands (Wiroatmodjo and Judi D.M., 1979). These forest areas are distributed in the various islands of the Indonesian archipelago (Table 1).

Flora and fauna

Mangrove forests became centers of interest for many biologists in the last century. In part, this interest was due to the unique form of the aerial roots of a mangrove tree. However, most of the studies carried out in mangrove forests were only to enumerate the tree species. Table 2 is taken from Kartawinata *et al.* (1979), which in turn have improved the listing of flora species by Backer (1911), van Steenis (1935, 1958) and Soemodihardjo *et al.* (1977). It is understandable that the species composition varies from one forest to another and from one island to another.

Ecologically, the mangrove forests represent a rather sharp transitional gradient between the ma-

Table 1. Areal distributions of mangrove forests in Indonesia*

I.	Sumatra:	Aceh	50.0	
		North Sumatra	60.0	
		Riau	75.0	
		South Sumatra	195.0	
		Lampung	17.0	
		Total Sumatra		397.0
II.	Java:	West Java	20.4	
		Central Java	14.0	
		East Java	6.0	
		Total Java		40.4
III.	Kalimantan:	East Kalimantan**	20.0	
		South Kalimantan	20.0	
		Central Kalimantan	10.0	
		West Kalimantan	40.0	
		Total Kalimantan		90.0
IV.	Celebes:	South Celebes	24.0	
		Southeast Celebes	29.0	
		Total Celebes		53.0
V.	Molluccas:	(North and Central Halmahera, Aru, Buru, Taliabu)		100.0
VI.	Irian Jaya:	(South coasts and North coasts)		2,934.0
VII.	Nusa Tenggara:	(Sumbawa and Timor)		3.7
		Total for Indonesia		3,627.1

* Preliminary Data after P. Wiroatmodjo and Judi D.M., 1978, in 'thousands of hectares'.
** The latest data show 120,000 ha.

Table 2. List of flora found in a mangrove forest (after Kartawinata, *et al*, 1979).

WOODY TREES

Apocynaceae
Cerbera mangnas l.
Bignoniaceae
Dolichandrone spathacea (l.f.) K. SCHUM
Combretaceae
Lumnitzera littorea (JACK) VOIGT
L. lutea PRFSI
L. rasemosa WHID
Euphorblaceae
Excoecaria agallocha L.
Leguininosae
Cumingla philippinensis VIDAL.
Cynometra ramiflora L.
Pithecellobium umbellanum (VAIL.) BRH.
Myrsinaceae
Aegiceras corniculatum (l.,) BLANCO
A. floridum R. & S.
Myrtaceae
Osbornea octodonta F.v.M.
Palinae
Nypa fruticans WURMB.
Oncosperma tigillaria (JACK) RIDI.
Phoenix paludosa ROXB.
Rhizophoracae
Bruguiera cylindrica (l.,) l.MK
B. exaristata DONG HOU
B. gymnorrhiza (L.) LMK
B. sexangula (l.our.) POIR.
Ceriops decandra (GRIFF.) DING HOU
C. tagal (PERR.) C.B. ROBINS.
Kandelia candel (L.) DRUCE
Rhizophora apiculata BI.
R. mucronata LMK.
R. stylosa GRIFF
Rubiaceae
Scyphiphora hydrophyllacea GAERN.
Rutaeae
Paramignya angulata (WILLD.) KURZ
Sonneratiaceae
Sonneratia alba J.E. SMITH.
S. caseolaris (l.) ENGL.
S. ovata BACK
Sterculliaceae
Heritiera littoralis DRYND. ex W. ALL.
Verbenaceae
Avicennia alba HI.
A. marina (FONSK.) VIRGII
A. officinalis l.

SHRUBS

Leguminosae
Caesalpinia crista l
Rubiaceae
Ixora timoriensis DIENI
Tillaceae
Brownlowia argentea KURZ
B. lanceolata BIH.
Verbenaceae
Clerodendrum inerme (L.) GAERTN.

HERBS AND GRASSES

Acanthaceae
Acanthus ebracteatus VAHI.
A. ilicifolius I.
A. volubilis WALL.
Araceae
Cryptocoryne ciliata (Roxb.) Scholl
Cyperaceae
Cyperus malaccensis LAMK.
Diplachne jusca (I.) BLAUV.
Fimbrystilis ferruginea (L.) VAHL.
Pteridaceae
Acrostichum aureum I.
Gramineae
Aegialitis annulata R. BR.

LIANA

Asclepiadaceae
Cynancluon carnosum SCHIRR.
Finlaysonia obovata WATT.
Sarcolobus banksii R. & S.
Gymnanthera paludosa (BT.) K. SCHUM.
Leguminosae
Dalbergia caudenatensis (DENNST.) PRAINS.
D. menoeides PRAIN.
Derris heptaphylla (L.) MERR.
D. heterophylla (WILLD.) BACKER
Rhamnaceae
Smythea lanceata (Tul.) Summerü.

EPIPHYTES

Adianthaceae
Vittaria sp.
Asclepiadaceae
Dischidia benghalensis COLE BR.
Hoya sp.
Davalliaceae
Davallia sp.
Ericaceae
Rhododendron brookeanum LOW.

71

Table 2. (continued).

Jungermanniceae
  Frullania sp.
Melastomataceae
  Pachycentria constricta (BT.) BI.
  Plethiandra sessilifolia RIDL.
Orchidaceae
  Aerides odorata LOUR.
  Anota violacea SCHIRR.
  Bulbophyllum xylocarpi J.J.S.
  Dendrobium alvifolium (BL.) RCHB.f.
  D. callibotrys KIDL.
  D. prostratum RIDL.
  D. rhizophoreti J.J.S.
  D. subulatum (BL). LINDL.
  Oberonia lacta J.J.S.
  O. rhizophoreti J.J.S.
Polypodiaceae
  Cyclophorus cinnamomeus V.A.V.R.

Drynaria sparsisora (DESV.) MOORE
Drymoglossum heterophyllum C. CHR.
Humata parvula (WALL.) MFTT.
Nephrolepis acutifolia (DESV.) VHR.
Platycerium coronarium (KOENIG) DESV.
Phymatodes scolopendria (BURM.) CHING.
P. simeosa (WALL.) J. SM.
Rubiaceae
  Hydnophytum formicarum JACK
  Myrmecodia sp.
Schyzaeaceae
  Lygodium lasum PR.

PARASITES
Loranthaceae
  Amyema gravis DAVIS.
  Viscum orientalis. I.

rine and the freshwater environments. Therefore, only flora and fauna that have strong tolerance in that kind of environment can survive. This factor more or less predetermines the number of species that exist in a mangrove ecosystem. Each biota or group of biota occupies a niche and even forms an aggregate and spatial zonation of its own. Van Steenis (1958) mentioned the combination of factors that may cause the species ecological preference such as:

- soil types: hard or soft texture, ratio between the sand and mud contents
- salinity: daily variation, annual average, the length of time, depth and how frequently the ground is submerged
- species tolerance of current and wave action
- general tolerance of their young (seedlings, larvae) for the preceding three factors

The mangrove fauna consists of terrestrial components and marine components. Examples of the terrestrial components are: bat (*Pteropus vampyrus*), varanus (*Varanus salvator*), wild pig (*Sius* sp), monkey (*Maccacairus, Nassalis larvatus*), various birds and insects. The marine component fauna generally is more prevalent than the terrestrial and tend to be dominated by gastropods and brachyurans (Table 3).

The first relatively complete record of fauna in an Indonesian mangrove forest was probably that of Verwey (1929) from the Jakarta Bay area. He listed various groups of animals, including mammals, reptiles, birds, amphibians and insects. In addition he (Verwey, 1930) made a special study on the distribution of crabs. Based on his investigations he distinguished various zonations based on the dominant species of crab, *e.g.* a *Sesarma taeniolata* zone, a *Uca consobrinus* zone, a *Uca signatus* zone, a *Metaplex elegans* zone and a *Scylla serrata* zone. Other studies have concentrated only on certain group of animals.

In addition to the sedentary fauna, there are a number of species that utilize the mangrove ecosystems only as temporary habitat, whether it be for spawning, for nursery, or only for shelter. For example, many economic species of shrimp have been shown to be mangrove dependent (Macnae 1974; Unar 1972). Therefore, the shrimp fisheries in Indonesia normally are strongly correlated with the distribution of mangrove forests (Martosubroto, and Naamin, 1977; Martosubroto, 1979).

Table 3. List of marine fauna found in a mangrove forest (after Kartawinata, *et al*, 1979)

GASTROPODA

Potamididae
    Terebralia palustris (LINNAEUS)
    T. sulcata (BORN)
    Telescopium telescopium LINNAEUS
    T. mauritsi BUTOR
    Cerithidea djadjarensis (MARTIN)
    C. alata (PHILIPPE)
    C. obtusa (LAMARCK)
    C. quadrata SOWERRY
    C. weyersi DAUTZEMBERG
    C. cingulata (GMFLIN)

Ellobiidae
    Cassidula aurisfelis Bruguiere
    C. lutescens BULOT
    C. mustelina DISCHAYIS
    C. triparietalis (MARTINS)
    C. sulculosa (MUSSON)
    Auriculastra subula (QUOY & GAIMARD)
    Ellobium aurisfudac LINNAEUS
    E. aurismidae (LINNAEUS)
    E. tomatelliforme
    Pythia plicata (FERUSSAC)
    P. trigona (TROSCHLI)
    Melampus singaporensis (PFEIFF)
    M. pulchellus
    M. semibuleatus MOUSSON

Littorinidae
    Littorina scabra (LINNAEUS)
    L. carniflora (MENKE)
    L. intermedia PHILIPPI
    L. melanostoma GRAY
    L. undata GRAY

Neritidae
    Nerita planospira ANTON
    N. albicilla LINNAEUS
    Neritina violacea (GMFLIN)
    N. turrita (GMFLIN)
    N. lineata (LAMARCK)
    N. zigzag LAMARCK
    N. varlegata LESSON
    N. auriculata LAMARCK
    Clithon corona (LINNAEUS)

Thiaridae
    Melanoides riqueti (GRATELOUP)
    M. tuberculata (MÜLLER)

Amphibolidae
    Salinator humana (BLANFORD)
    S. fragilis (LAMARCK)

Cerithidae
    Cerithium morum LAMARCK

Melongenidae
    Melongena galeodes LAMARCK

Trochidae
    Monodonta labio (LINNAEUS)

Assimineidae
    Syncera brevicula (PFEIFFER)
    S. javana ((T HILLER)
    S. nitkia (PEASE)
    S. woodmasonlana (NEVILLE)

Stenothyridae
    Stenothyra glabrata (A. ADAMS)

BIVALVIA

Larbiculidae
    Polymesoda coaxans GMELIN
    P. expansa (MOUSSON)

Veneridae
    Gafrarium tumidum RODING

Anomiidae
    Enigmonia aenigmatica (CHEMNITZ)

Ostreidae
    Ostrea cucullata BORN

Arcidae
    Anadara antiquata LINNAEUS

CRUSTACEA

Grapsidae
    Sesarma ideniolata WHITE
    S. meinerti (DE MAN)
    S. bataviana (DE MAN)
    S. cumolpe DE MAN
    S. smithi H. Milne – EDWARDS
    S. bocourti A. MILNE – EDWARDS
    S. fasciata LANCHESTER
    S. bidens DE HAAN

Table 3. (continued).

S. onychophora DE MAN
S. rousseauxi H. MILNE – EDWARDS
S. erythrodactylum HESS
S. longipes (KRAUSS)
Metopograpsus latifrons (WHITE)

Ocypodidae
   Uca vocans LINNAEUS
   U. lactea (DE HAAN)
   U. triangularis A. MILNE – EDWARDS
   U. signatus (HESS)
   U. consobrinus (DE MAN)
   U. annulipes (H. MILNE – EDWARDS)
   U. dussumieri (H. MILNE – EDWARDS)
   Macrophthalmus convexus STIMPSON
   M. telescopicus OWEN
   M. definitus ADAM & WHITE
   Ocypoda ceratophthalma (PALLAS)
   O. arenaria DE MAN
   Ilyoplax delsmani DE MAN

   I. orientalis (DE MAN)

Portunidae
   Scylla serrata (FORSKAL)

Gegarcinidae
   Cardisoma carniflex (HERBST)

Thalassinidae
   Thalassina anomala HERBST

Alpheidae
   Alpheus crassimanus HELLER
   A. bisincisus DE HAAN

Paguridae
   Coenobita cavipes STIMPSON

Balanidae
   Balanus sp.

---

c. Zonation

In addition to species composition one can differentiate various zones. These zones are governed by the interaction of factors such as the frequency and length of time of submergence, the salinity and the drainage system. Generally each zone is called by the dominant woody tree species, for example there can sometimes be distinguished (from outer edge to the inner zone): a *Sonneratia* zone, an *Avicennia* zone, a *Rhizophora* zone, a *Bruguiera* zone, a *Ceriops* and a *Nipah* (*Nypa fruticans*) association.

In this classic study on the effect of submergence and salinity on the species composition of the mangrove trees in Cilacap, Haan (1935) distinguished six different zones, as follows:

Class 1:

range of salinity 10–30‰; The ground is submerged once or twice daily at least 20 days per month; the species of *Avicennia* or *Sonneratia* on new soft ground or *Rhizophora* on hard ground; to form the outer zone.

Class 2:

range of salinity 10–30‰, The ground is submerged 10–19 days per month; *Bruguiera gymnorrhiza* grows well; forming the middle zone.

Class 3:

range of salinity 10–30‰ the ground is submerged 9 days or less per month; species of *Xylocarpus* and *Heritiera* grow here; forming the third zone.

Class 4:

range of salinity 10–30‰; the ground is submerged a few days in a year only, species of *Bruguiera*, *Scyphiphora* and *Lumnitzera* grow well in this inner zone.

Class 5:

salinity 0‰; the ground is affected very little by tides.

Class 6:

salinity 0‰; the ground is affected by water level only during the wet season.

Class 5 and class 6 represent the transitional zones to the freshwater swamps behind the mangrove forest. In this zone grow species of *Cerbera* and *Oncosperma*.

Verwey (1930) on the other hand had proposed another type of mangrove zonation based on the dominant species of crabs that inhibit the area. He had identified five different zones: a *Sesarma taeniolata* zone, a *Uca consobrinus* zone, a *Uca signatus* zone, a *Metaplex elegans* and *Scylla serrata* zone.

d. Physical and chemical properties

Studies on the physical and chemical properties of the mangrove ecosystem in Indonesia are still few. Most of the studies were carried out in attempts to determine the suitability of mangrove forest for conversion into rice fields (e.g. IPB, 1969 and 1975; Van Wijk, 1951). Notohadiprodjo (1979) classified the stability of mangrove soil aggregates based on the percentage of organic carbon as follows:

| Aggregates | % C organic | |
	Ranges	Average
very unstable	2.86–10.96	7.60
unstable	0.97–21.79	4.23
less stable	2.00–11.80	6.53
little less stable	2.83–14.33	8.69
stable	5.55–12.10	9.37

From his samples of soil texture taken from the mangrove forest in Cilacap he found out that:

91% of the soil contained 40% or more clay

65% of the soil contained 60% or more clay

96% of the soil contained higher percentage of colloidal clay (less than 0.5 micron in diameter).

53% of the soil contained higher percentage of

Table 4. Chemical and physical preperties of the soil from three mangrove communities in the Rambut Island Nature Reserve, Jakarta Bay (after Kartawinata and Walujo, 1977).

Soil properties	Scyphiphora pemphis community	R. mucronata community	R. mucronata R. stylosa community
Soildepth (cm)	0–10	0–50	0–10
Texture (%)			
Sand	71.5	11.3	90.0
Silt	17.5	44.7	6.0
Clay	11.0	44.0	4.0
pH	7.6	6.7	7.6
Organic matter (°)	7.8	62.1	2.0
Nutrients (°)			
N	0.28	1.26	0.10
P_2O_5	0.282	0.075	0.054
K_2O	0.082	0.283	0.040
CaO	47.962	11.357	49.723
-MgO	0.776	1.704	0.874
Exchangeable cations (m.e.)			
Ca	33.9	68.8	27.1
Mg	14.2	69.5	7.5
K	1.3	6.4	0.8
Na	29.5	149.3	21.4
Total	79.1	292.4	56.8
Adsorption capacity	23.5	84.4	5.6

medium silt than other fractions.

Kartawinata and Walujo (1977) have made some studies on the physical and chemical properties of the soil in three different mangrove communities in the Rambut Island nature reserve, Jakarta Bay (Table 4).

e. Utilizations

For centuries the Indonesian people traditionally have utilized mangroves, mostly for firewood, charcoal, tannin, dyes, timber and even boats. The following genera are frequently used for those purposes: *Rhizophora, Bruguiera, Ceriops, Avicennia, Nypa* and *Oncosperma*. The nypa leaves can be used for thatch roof, baskets, and cigaret paper.

The stalks of the flower can be cut and the sap used for making brown sugar or arak (a kind of liquor). The Riau archipelago has long been known as a center for mangrove charcoal. The mangrove products mostly are exported to Singapore and Hong Kong. The Forestry Service of the Riau Archipelago has reported that in Riau there are about 556 charcoal producers with an average production of 25 tons of charcoal per burning. On the average each producer can burn charcoal six times in one year. The following table gives some idea as to the volume of production of charcoal and other products from the Riau archipelago:

Production of Riau mangroves*

Year	Charcoal (in tons)	Firewood (in m³)	Pulpwood (in m³)
1973	20,079	21,715	22,038
1974	22,322	52,852	39,412
1975	23,385	52,261	63,347
1976	27,251	12,063	63,347

* Forestry Service of Riau Province

In recent years mangrove wood has been used also as raw material for a paper mill in Gowa, south Celebes. The Gowa paper mill used bamboo as the basic raw material and mangrove wood as a mixture. They found out that by processing with a ratio of 80% bamboo and 20% mangrove they could obtain a very good quality paper (Rachmat, 1975). In 1974 Chipdeco company started to produce chipwood from mangrove in East Kalimantan. This company exploits about 85.000 hectares of mangrove forest in East Kalimantan.

Some environment and management problems

The mangrove ecosystem has many functions. As has been outlined above the mangrove forest can be exploited for the firewood, charcoal, timber and for raw materials for paper and chipboard. In addition, it can function also in coast stabilizer, erosion prevention, as spawning and nursery ground, as food sources of many groups of animals and even as a pollutant trap. Most of the environmental problems in a mangrove ecosystem usually are related to its mode of exploitation and to its various ecological function.

For example, Versteegh (1952) mentioned that the exploitation of mangrove forest in Indonesia generally is not based on conservation principles. He pointed out the wide spread practice of cutting down mangrove trees for many years in the Riau archipelago, without any conservation measures. For a charcoal production he designed a 30 year exploitation cycle as follows: the mangrove forest is divided into many sub-blocks of 120 ha each. Only 4 ha of forest should be cut from each sub-block. Thus the exploitation cycle is 120 ÷ 4 = 30 years. In order to ensure a long term exploitation, at present many forest companies follow Versteegh's management scheme. The exploitation cycle may vary from 20 to 40 years, depending on the density of the stands and area of the forest to be exploited.

There was only one study on the impact of pollution specifically carried out in a mangrove ecosystem, *i.e.* the study carried out in the Cilacap mangrove ecosystem by Hardjosuwarno *et al.* (1975). The Cilacap area has been designated as an industrial complex for the southern coast of central Java. An oil refinery has operated since 1977 in the area. In addition in Cilacap one can find ferro-sand mining, an oil terminal and a fishing port as well as an open ocean port. This complex is situated adjacent to a 23,000 ha mangrove forest, the most extensive mangrove forest in Java. In the study, the following physical and chemical parameters were monitored: temperature, pH, DO, CO_2, carbonate contents, alkalinity and hydrocarbon. Although, the hydrocarbon contents of both the mud and seedlings were very high, on the average 770 ppm and 6150 mg/seedlings respectively, but the authors were unable to draw any conclusion on the effect of petroleum pollutant on the mangrove ecosystem nor to the productivity of the water around it.

In order to preserve the ecological functioning of mangrove forest the Indonesian government, through the Department of Agriculture and Directorate General of Forestry, has regulated the cutting of mangrove forest. It requires that 50–200 meter wide belt of mangroves be retained along the coast. This belt serves not only to preserve the ecological functioning of mangrove forest, but also to ensure the natural regeneration process in the region. In addition, in order to ensure that natural ecosystems will still exist for future studies and future development, a number of nature reserves have been established in Indonesia. Many of these reserves are located in the coastal areas. Currently the Indonesian government is exploring the possibility of increasing the nature conservation area, from the present 3.5 million hectares to 10 million hectares by 1984. A marine park and reserve system is planned. Many potential sites, including mangrove and coral reef ecosystems, have been identified. Surveys and studies on these sites will be carried out jointly by the Indonesian National Institute of Oceanology, the Directorate of Nature Protection and Conservation and the World Wildlife Fund.

However, probably the most urgent problem facing us concerning the mangrove ecosystem is the lack of data and knowledge on which to base management and development schemes. The followings are some of the suggested research programs for mangroves:

a. Inventory and detail of areal coverage.

b. Species composition and species diversity for both the flora and fauna and the zonation based on this species diversity.

c. The food chains, biological aggregations and the energy flow in various communities or trophic levels.

d. The physical and chemical properties of the soil and the water in and around the ecosystem.

e. The biology of reproduction of mangroves, including the methods of pollination and dispersion.

f. Relation between a mangrove ecosystem and fisheries in the surrounding waters.

g. Study on the impact of some external pressures, e.g. pollution, over harvesting, conversion to other uses etc., on the survival and productivity of mangrove ecosystem.

h. Silviculture practices in order to obtain the maximum sustainable yield.

Certainly, there is no single country that is able to carry out these programs, at least in part due to the inadequate number of skilled and trained manpower. Therefore, bilateral regional and international co-operation is being developed for this purpose.

Literature cited

Backer, C.A. 1911. Schoolflora van Java. Weltevreden, 478 pp.

Haan, J.H. de. 1931. De Tjilatjapsche Vloedboschen (The tidal forest of Cilacap). Tectona 24:39–76.

Hardjosuwarno, S., S.D. Tandjung, A. Sukahar, A. Pudjoarinto and Purwoto. 1975. Studi mengenai ekosistem dari mangrove community Cilacap (Jawa Tengah) A study on the mangrove ecosystem of Cilacap (Central Java). Report, Faculty of Biology, Gadjahmada University, 42 pp.

IPB (Bogor Agriculture University). 1969. Preliminary report on a survey of tidal irrigation areas in the provinces of Riau, Jambi and South Sumatra. A Preliminary Report.

IPB 1975. Ecological survey on Upang Delta region. Natural resources aspects and their management, Book I–IV. A Preliminary Report.

Kartawinata, K., S. Adisumarto, S. Soemodihardjo and I.G.M. Tantra, 1979. Status pengetahuan hutan bakau Indonesia (The state of knowledge on the Indonesian mangrove forest). Proceeding of Seminar on the Mangrove Ecosystem, LHL 26:21–39.

Kartawinata, K. and E.B. Walujo. 1977. A preliminary study of the mangrove forest on Pulau Rambut, Jakarta Bay. Mar. Res. Indon. 18:119–129.

Martosubroto, P. 1979. Sumbangan hutan mangrove terhadap perikanan (contribution of mangrove forest to fisheries). Proceeding of the Seminar on Mangrove Ecosystem. LHL 26:109–113.

Martosubroto, P. and N. Naamin. 1977. Relationship between tidal forest (mangrove) and commercial shrimp production in Indonesia: Mar. Res. Indon. 18:81–86.

Macnae, W. 1974. Mangrove forests and fisheries. FAO/IOFC/ DEV/74/34. FAO, Rome, 35 pp.

Notohadiprodjo, T. 1979. Beberapa sifat tanah mangrove ditin-jau dari segi edafologi (Some features of mangrove soil as viewed from the edaphologic aspect). Proceeding of Seminar on Mangrove Ecosystem. LHL 26:40–54.

Rachmat, M. 1975. Pentrapan high yield industrial plywood plantation di hutan Borisallo, Perum Kertas Gowa (The application of high yield industrial plywood plantation in Borisallo forest, Gowa Paper Mill Company). Kehutanan Indonesia:675–680.

Soegiarto, Aprilani. 1980. Status report of research and monitoring of the impact of pollution on mangrove and its productivity in Indonesia. UNEP/FAO Regional Southeast Asian Consultative Meeting on Mangrove Ecosystem, Manila, 4–9 February, 1980, 51 pp + Appendix.

Soemodihardjo, S., K. Kartawinata and S. Prawiroatmodjo. 1977. Kondisi hutan payau di Teluk Jakarta dan pulau-pulau sekitarnya (The condition of mangrove forests in the Jakarta Bay and its surrounding islands). OSEANOLOGI DI INDO-NESIA 7:1–23.

Steenis, C.G.G.J. van. 1935. Maleische vegetatischetsen (The Malaysian vegetation). Tijdschr. Kon. Ned. Aardr. Gen. 52:25–67, 171–203, 363–398.

Steenis, C.G.G.J. van. 1958. Ecology. An introductory part to the monograph of Rhizophoraceae By Ding Hou. Flora Maleisiana 5:431–441.

Unar, M. 1972. Review of the Indonesian shrimp fishery and its present development. publ. Mar. Fish. Inst. (LPPL) 1/72:1–26.

Unar, M. 1979. Survai udang di perairan Teluk Waworada dan pantai selatan Timor (Prawn surveys in Waworada Bay and south coast of Timor). Proceeding of the Seminar on Mangrove Ecosystem. LHL 26:205–212.

Van Wijk, C.L. 1951. Soil survey on the tidal swamps of South Borneo in connection with the agricultural possibilities. Contr. Gen. Agric. Res. Sta., Bogor No. 123:49 pp.

Versteegh, F. 1952. Problem of silviculture and management of mangrove forests in Indonesia. Paper presented at the Asia Pacific Forestry Commission Conference, Singapore, Dec. 1–13.

Verwey, J. 1929. The coral reefs in the Bay of Batavia. Part II. Zoology. Fourth Pacific Science Congress, Bandung.

Verwey, J. 1930. Einiges über die biologie Ost-Indische Mangrove Krabben. Treubia 12:169–261.

CHAPTER 11

Mangrove management in the Philippines

AIDA R. LIBRERO

Socio-Economics Research Division, Philippine Council for Agriculture and Resources Research and Department of Agricultural Economics University of the Philippines at Los Baños

Introduction

The Philippines is an archipelago composed approximately 7,100 islands stretching more than a thousand miles in length. It is bounded in the east by the Pacific Ocean and in the west by the China Sea. Bridging the gap between the land and the sea are the wetlands which blend both the aquatic and terrestrial ecosystems. Such wetlands are composed of mangroves which serve as nursery grounds for many organisms. They are the sources of food and nutrients for aquatic life and therefore would affect the productivity of marine resources. In addition, mangrove areas have been known to retard soil erosion and flooding. Likewise, the forest cover provides housing materials, firewood, and charcoal. There are therefore, competing demands for this resource: the forester wants to utilize the trees; the fish farmer wants to convert it into fishponds; and the environmentalist wants to conserve the area for breeding and ecological stabilization. (Proceedings, 1977).

The implications of the fate of mangrove forests for food production and for the social and economic life in coastal communities are far-reaching. Massive pollution, wasteful consumption, uncontrolled resource use can result in disastrous climatic changes, diseases, and food shortages.

Mangrove resources

The Philippines Bureau of Forest Development (BFD) estimated that the Philippines in 1970 and a total of 246,699 hectares of mangrove forest, equivalent to 1.9% of the total forest area (Table 1). Of this total 223,771 ha are forest land and 22,928 ha are alienable and disposable. The mangrove forests are geographically distributed as follows: Mindaneo – 46.8%; Visayas – 26.9%; Palawan – 16.4%; and Luzon – 9.9%. These data were based on the definition of mangrove as a 'forest type consisting of forest stands found in the swampy tidal areas composed primarily of *Rhizophora* and associated species.' These were obtained or compiled through regional surveys and interpretations of aerial photographs in some cases.

The Bureau of Fisheries and Aquatic Resources (BFAR), on the other hand does not limit mangrove areas to the forest stands but they include all intertidal areas. They even extend to areas where the 'nature of the soil although without mangrove trees are identically of parallel origin and the ecological conditions therein are suited for aquaculture.' Altogether, this definition covers the total mangrove ecosystem below the high tide mark. Statistics from BFAR therefore revealed a quite different estimate for the country's total mangrove area compared with that of BFD. However in 1968 BFAR started using the same data from the BFD so that now both have similar estimates for total mangrove arease, altough BFAR still considers the

H.J. Teas (ed), Physiology and management of mangroves.
© *1984 Dr W. Junk Publishers, The Hague. ISBN 90 6193 949 6. Printed in the Netherlands.*

Table 1. Mangrove forest of the Philippines by geographical region, 1978 (hectares).

	Forest Land	Alienable and Disposable	Total
Philippines	223,771	22,928	246,699
Luzon	22,532	1,977	24,509
Palawan	33,981	6,349	40,330
Visayas	58,734	7,627	66,361
Mindanao	108,524	6,975	115,499

Source: Philippines, Bureau of Forest Development, 1978 Philippine Forestry Statistics.

non-forest 'marshes and small water' (BFD classification) as suited for fishpond development.

Mangrove areas are a fast dwindling resource. While mangrove forest may be classified academically as a renewable resource, the current practices of exploitation and utilization result in them being non-renewable. For the period 1967–1976, mangrove areas in the Philippines declined from 418,990 ha to 249,138 ha or approximately 16,741 ha annually (Table 2). These areas were cleared for fishpond or simply due to the extraction of fuelwood and timber. The rate of decline has abated. The decline from 1977 to 1978 was 2,439 ha.

A forest area, whether upland or mangrove can be classified according to its stand size as follows:

a. Reproduction brush – productive forest stands predominantly stocked with tree reproduction or brush, i.e. trees at least one meter high but less than 15 cms in diameter.

b. Young growth – productive forest stands predominantly stocked with young trees 15 cms. or larger in diameter. Most stands in this class have been cut over with residual trees remaining. Stands stocked mainly with mature trees but with 25% or more of mature stand volume removed by cutting qualify as young growth also.

c. Old growth – productive forest stands predominantly stocked with mature tree with less then 25% of the mature stand volume cut.

According to stand size, 50.16% of mangrove forests in the Philippines are reproduction-brush (Table 3). Majority of this type are in the Visayas. Young growth stands accounted for 45.08%, mostly located in Mindanao. A small portion (4.76%) are of old growth type (4.76%) all of them found in Mindanao and Palawan.

Utilization

The importance and varied uses of mangroves areas could not be overemphasized especially in developing countries. From the ecological viewpoint, mangroves serve as an effective buffer to the upland environment during storms by preventing soil erosion and minimizing sea water pollution by

Table 2. Distribution of mangrove forest by geographic region 1969–1978 (hectares).

Year	Total	Region			
		Luzon	Palawan	Visayas	Mindanao
1969	295,190	42,520	47,534	77,479	127,657
1970	288,035	42,204	44,426	77,825	123,580
1971	286,640	42,084	44,421	77,469	122,676
1972	284,211	40,794	43,938	77,262	122,217
1973	258,895	30,969	42,746	67,386	117,794
1974	256,456	29,669	42,263	67,189	117,335
1975	254,016	28,199	41,779	66,928	116,876
1976	251,577	27,089	41,396	66,775	116,417
1977	249,138	25,799	40,813	66,568	115,959
1978	246,699	24,509	40,330	66,361	115,499

Source: Philippine Forestry Statistics, 1968–1978

serving as coastal pollutant sink or trap. They also act as shelter and habitat for wildlife. They also aid in building land by accumulating sediments.

In forestry, the significance of mangrove lies in the timber and non-timber products it provides. The different products derived from them include: timber for construction and furniture materials, viscose rayon for textils fibers, tannins for hide in leather manufacture, firewood and charcoal for fuel, nipa thatch and shingles for cottage industries, roofs and walls of huts, sap from palms for vinegar and wine making.

There are at present 8 species of timber (*Rhizophora* spp. and *Xylocarpus granatum*) and one species of palm (*Nypa fruticans*) found in mangrove forests being utilized for economic purposes. As of 1978, the volume of standing timber in the mangrove forest of the Philippines was 7,468 thousand m³ accounting for 0.46% of the total volume of standing timber in all forest area. Of this total, young growth stands accounted for 64% (Table 4).

Geographically, Mindanao has the largest volume of standing timber followed by Palawan (Table 5). The same table shows the declining trend in the volume of standing timber. However, the exploitation of mangrove forests has persisted without reforestation or regeneration to restore the original vegetation.

The extent of utilization of the mangrove forest for non-timber products is manifested in their production through the years (Table 6). From 1971 to 1977, 24,469 m³ of charcoal were produced from mangroves. This was one-fourth of the total charcoal production from forests; upland forest sources accounted for three-fourths.

Firewood production in 1979 amounted to 75,339 m³ of which 31% came from mangroves.

For fisheries, mangrove areas serve as breeding, spawning and nursery grounds for numerous species of fishes, molluscs and crustaceans. The tidal characteristics of the mangrove swamps lead to the suitability of this ecosystem for aquaculture purposes. The tide brings in the nutrient-rich water of the estuary and draws out the undesirable water in the fishponds during ebb-tides. For both fishermen and fishfarmers, therefore, this piece of ecosystem is very important.

The suitability of mangrove areas for fish production encouraged both public and private entities to tap these areas for fishpond development.

Table 3. Distribution of mangrove forest area by stand size and by geographical region, 1978 (hectares).

Stand Size	Philippines	Luzon	Visayas	Mindanao	Palawan
Reproduction brush	123,746	24,509	66,361	25,958	40,330
Young growth	111,212	10,795	64,987	83,978	22,006
Old growth	11,741	–	–	5,563	12,146
Total	246,699	35,294	131,348	115,499	74,482

Source: 1978 Philippine Forestry Statistics.

Table 4. Volume of standing timber in mangrove forest by stand size and geographical region, 1978 (thousand cubic meters).

Stand Size	Philippines	Luzon	Palawan	Visayas	Mindanao
Reproductive brush	1,198	71	402	512	213
Young growth	4,762	482	1,452	28	2,800
Old growth	1,508	–	1,062	–	446
Total	7,468	553	2,916	540	3,459

Source: Bureau of Forest Development. 1978 Philippine Forestry Statistics.

Table 5. Volume of standing timber in mangrove forest land by geographical region (000 m³).

Year	All forest area	Region				
		Philippines	Luzon	Palawan	Visayas	Mindanao
1970	1,937,048	9,777	826	4,245	728	3,978
1971	1,967,703	9,739	824	4,244	725	3,947
1972	1,768,367	9,513	747	4,136	669	3,977
1973	1,682,058	8,077	569	3,364	543	3,602
1974	1,690,829	8,920	670	3,414	837	3,998
1975	1,695,532	10,046	[a]	4,357[a]	1,298	4,391
1976	1,648,443	9,647	764	3,468	1,052	4,363
1977	1,660,547	7,526	553	2,993	536	3,444
1978	1,623,587	7,468	553	2,916	540	3,459

[a] Combined value was for Luzon and Palawan.
Source: Philippine Forestry Statistics, 1970–1978.

Table 6. Production of some minor forest products from mangrove forest, 1971–1977.

Year	Charcoal (m³)	Firewood (m³)	Nipa		Tanbark (kilos)
			Shingle (kilo)	Sap (Liter)	
1971	6,833	118,152	1,473,088	56,800	312,998
1972	2,308	23,752	981,300	24,500	322,200
1973	3,759	3,710	902,975	11,600	393,950
1974	2,494	14,433	683,600	202,000	589,205
1975	4,585	7,488	1,048,830	11,950	822,600
1976	2,791	9,190	629,650	11,950	785,286
1977	1,699	23,188	736,281	25,000	131,150

Source: Philippine Forestry Statistics 1971–1977.

In the early 1950's area devoted to fishponds in the Philippines was only about 88,681 hectares. This went up to 100,097 ha in 1954, further increasing to 129,062 ha in 1962 (Table 7). Ten years later, i.e. in 1972, fishpond area was 174,101. The area stabilized at about 176,000 hectares in the 1970's. However, as of 1978 the National Mangrove Commitee reported 2,795 applicants for fishponds covering 52,393 ha. This area is roughly 21% of the remaining mangrove area of the Philippines.

The magnitude of the conversion of mangrove areas into agricultural production, human settlements, industrial development and other uses is not documented. However, observations have indicated that extensive areas that have been utilized for such purposes.

Polices/regulations governing the development. Preservation and/or conservation of the mangrove areas

The government has adopted a general policy of protection, conservation and wise utilization of mangrove resources because of its recognition of the ecological and economic importance of mangrove areas both for the present and the future generations.

At present there is no single agency responsible for the management and development of mangrove areas. The differences in the sectoral approach to development, that is, for forestry, fishing, fish-farming, urban and industrial development, are being resolved by the present institutional structure.

Table 7. Fishpond area in Operation

Year	Area (ha)
1952	88,681
1953	95,633
1954	100,097
1955	104,952
1956	109,799
1957	112,611
1958	116,546
1959	119,582
1960	123,252
1961	125,810
1962	129,062
1963	131,850
1964	134,242
1965	137,251
1966	138,968
1967	140,055
1968	162,807
1969	164,414
1970	168,118
1971	171,446
1972	174,101
1973	176,032
1974	176,032
1975	176,032
1976	176,234
1977	176,231

Source: BFAR Fisheries Statistics.

At the ministerial level, the Ministry of Natural Resources (MNR) is the principal government agency responsible for the development, management and conservation of mangrove resources. Under the MNR are agencies whose functions relate to the disposition of mangrove areas. These are the Bureau of Forest Development, the Bureau of Fisheries and Aquatic Resources (BFAR), and the Bureau of Lands.

Also attached to the MNR are the Forest Research Institute (FORI), the Natural Resources Management Center (NRMC), the Fishery Industry Development Council (FIDC) and the National Environmental Protection Council (NEPC).

Other government agencies whose functions and jurisdiction relate to mangrove resources are the following: The National Pollution Control Commission (NPCC); the Philippine Coast Guard (PCG); the National Science Development Board (NSDB); the Philippine Council for Agriculture and Resources Research (PCARR); and various state colleges and universities.

Adopting a holistic approach in mangrove development, these agencies complement each other. They are mandated to implement policies and guidelines concerning the development, management and conservation of mangrove resource as embodied in several Presidential Decrees, Administrative and Special Orders.

a. Classification and survey

Policies on the classification and survey of mangrove areas date back to 1952 with Forestry Circular no. 95 formulated for the purpose of zonifying swamplands. Mangrove areas were to be zonified according to: (1) areas to be retained permanently for forest purpose; for the gathering of firewood, charcoal, nipa shingles, nipa sap and tanbark; for the construction of saltwork, stream bank protection and others; and (2) those areas that may be released for fishpond purposes. The zonification was not fully accomplished due to lack of funds and technical personnel to do the job (Philippine Lumberman, 1965).

In 1975, Presidential Decree No. 705, otherwise known as the Forestry Reform Code of the Philippines, designated the Ministry of Natural Resources to devise guidelines and methods for the proper and accurate classification and survey of all lands of the public domain into agricultural, industrial or commercial, residential, resettlement, mineral, timber or forest, and grazing lands.

The system of classification stipulated that mangrove and other swamps not needed for shore protection and suitable for fishpond purposes shall be released to and be placed under the administrative jurisdiction and management of the BFAR. For this purpose, Special Order No. 3 creating the Land Classification Composite Teams of the BFD was issued on July 22, 1975. Mangrove areas have been delimited by this team for such purposes as forest and fishpond development. This process of land classification is a vital factor in controlling conversion of mangrove areas into fishponds.

In another Special Order No. 309 of December 13, 1976, the National Mangrove Committee of the NRMC was created. This committee was charged with the areal assessment of the mangrove areas through remote sensing with the use of Land Satellites. In collaboration with BFD, BFAR, FORI, and FIDC, the Committee would undertake ground inventory and assessment of selected mangrove areas with the following expected outputs: (1) taxonomic research surveys on mangal fauna and flora; (2) qualitative studies on fish and shellfish resources, and (3) productivity and standing crop of mangrove resources.

b. Utilization

For purposes of utilization, P.D. No. 705 stipulates that an evaluation of the mangrove areas shall first be conducted before exploitation, utilization or occupation is allowed. It further specifies that optimum benefits should be derived from the exploitation of these resources. The issuance of licenses, leases and/or permits is required whether the resource will be used for fishpond development or for forest purposes.

For forest purposes, proper control in the utilization and exploitation is effected through the issuance of BFD of Ordinary Minor Timber License for the extraction of mangrove products for firewood and other construction materials, and Ordinary Minor Products License for nipa shingles and tanbarks and other similar products.

Among the requirements for license application are: a sketch map of the area applied for; evidence of capital investment of at least ₱1,000 per hectare; a business plan; a recommendation from an office concerned with cultural minorities, if the area falls within the area set for such purposes; a waiver of rights of existing licenses; a contract to supply NACIDA/PCHI/BOI registered manufacturer or processor; a copy of the NACIDA/PCHI/BOI certificate to be supplied; and payment of fees (application fee of ₱1.00 per hectare and an oath fee of ₱2.00). The license fee is equivalent to 5 percent of the allowable cut multiplied by the average forest charge per unit measure. A bond deposit of ₱1.00 per cu. m. in the case of bakawan (*Rhi-zophora* spp.) and ₱0.60 per thousand shingles in the case of nipa (*Nypa fruticans*) has to be posted.

For fishpond development, Presidential Decree No. 704 or the Fishery Decree maintains that the country's fishery resources be in optimum productive condition. Applicants wishing to lease swamplands for fishpond purposes are required by the BFAR to submit a project feasibility study to develop the area and make it productive at optimum levels. The government prohibits the disposal of public lands suitable for fishpond purposes by sale. These lands should only be leased to qualified persons, associations, cooperatives, or corporations. A qualified and interested applicant is granted a lease for a period of 25 years renewable for another 25 years. The lease contract gives him 5 years to develop 50% of the fishpond into commercial scale and another 5 years to develop in commercial scale the remaining portion. All areas not fully developed within this given period automatically revert back to the Bureau for disposition and is declared open/available for other applicants. The same holds if the leased area is abandoned or is not developed. Such management concept in effect shows that developing swamplands into fishponds must be done within a given period of time. Thus, land speculation is checked. The government aquaculture program, furthermore, concentrates on increasing the yield per hectare of existing ponds by introducing improved technology, rather than expanding the present hectarage to boost fish production. This is in recognition of the importance of the mangrove areas for other economic uses as well as the biological system of commercial fishes.

The average yield of fishponds in the Philippines is only about 600 kg/ha/yr but there are large differences between regions and farms within the same region. While some fishponds get only 100–200 kg/ha/yr, some farms get 1000 kg or more. Thus a substantial increase in production can potentially be achieved from the existing fishponds if better technology and management is adopted by producers.

While the development of mangroves into fishponds will increase fish production from aquaculture, it could also result in decreased production from coastal fisheries. Aquaculture provides 10%

of total fish supply and marine fisheries, 90%. The net result could be a reduction in total fish supply (Gabriel, 1978).

c. Development, preservation and/or conservation and management

Simultaneous to optimum utilization, proper management of mangrove resources is embodied in a number of decrees and administrative orders.

BFD Administrative Order No. 74, series of 1974 prescribes the employment of the Seed-tree-and-plant method of silvicultural treatment. This system of clear cutting leaves undamaged twenty or more seed tree per hectare with diameters of 20 cm or larger, well-scattered and strategically located in the forest area. Artificial regeneration or reforestation by the licensee concerned should augment this system, especially in open and heavily depleted areas.

Conservation-wise, P.D. No. 953 requires a holder of a lease agreement to plant trees extending at least 20 meters from the edge of river banks or creeks. Under the same decree, any person who cuts, destroys or injures naturally growing or planted trees of any kind in this area without authority from any government agency concerned is liable to a fine and/or imprisonment.

PD 705 also states that the following are needed for forest purposes and may not therefore be classified as alienable and disposable land: *a* Twenty-meter strips of land along the edge of the normal high water line of rivers and streams with channels of at least 20 meters wide, along shorelines facing oceans, lakes and other bodies of water, and strip of land at least 20 meters wide facing lakes; *b* strips of mangrove forests bordering numerous islands which protect the shoreline, the shoreline roads and even coastal communities from the destructive force of the sea during high winds and typhoons, and all mangrove swamps set aside for coast-protection purposes shall not be subjected to clearcutting operations.

Mangrove and other swamps released to the BFAR for fishpond purposes which are not utilized or which have been abandoned 5 years from the date of release shall not revert to the Category of

forest land. The National Mangrove Committee, has provided guidelines for the selection of mangrove areas to be preserved/conserved or to be declared as Mangrove Forest Reserved.

Based on ecological and socio-economic reasons, the mangrove areas to be recommended for preservation/conservation or to be declared as mangrove forest reserve are the following:

1. Mangrove areas adjoining mouth of major river systems

To maintain the ecological balance of estuarine areas, mangrove forest adjoining mouth of major river systems should be closed from fishpond development. The area to be preserved should cover at least a 3-kilometer stretch of mangrove on both sides of the mouth of the river fronting the sea.

2. Mangroves areas near or adjacent to traditional productive fry and fishing grounds

Considering the importance of mangroves as a breeding, spawning and nursery grounds for a variety of fishes and shellfishes mangroves near or adjacent to traditional productive fry and fishing grounds should not be alienated or released or fishpond purposes.

3. Mangrove areas near populated areas/urban centers

These mangrove areas should be conserved for the utilization of people who are dependent on mangrove forest products for their livelihood or domestic needs (i.e., firewood, wood for making charcoal, and timber for household construction).

4. Mangrove areas of significant hazard if developed, because of storms, erosion, floods, etc.

Mangrove forests which act as natural buffer against shore erosion, strong winds and storm floods should be left untouched.

5. Mangrove forests with primary/pristine and - dense young growth

Regardless of location, swamplands, which are

covered with virgin mangrove forest and dense young growth, should be preserved or declared as forest reserve because these are important in maintaining ecological balance in the mangrove ecosystem. These are also needed for riverbank and shore protection, wildlife sanctuaries, and for educational or research purposes.

6. *Mangrove forests on small islands*

Mangrove forests on small islands serve as major ecological components of the islands ecosystem and should in no case be disturbed.

Recommendations for research

During the National Symposium/Workshop on Mangrove Research and Development held in the Philippines in 1977 the following recommendations were made to provide guidelines for research and developmental projects.

Priority areas should be given to research areas that will provide basic information for more advanced investigation, development planning and management/conservation measures. Accurate data on the geographical distribution of mangrove swamps are needed. Likewise, a nationwide inventory of mangrove forests will show which areas are depleted or on the verge of depletion. Such inventory should lead to the evaluation of the economic potentials of mangrove areas. It will also help determine more comprehensively the mangal species present in the country, their relative abundance and volume.

Studies on both the mangal and faunal community structures and on species association and interrelationships will contribute to a better understanding of this unique ecosystem. Similarly, the characteristic physiochemical and geographical features of the mangrove ecosystem must be investigated.

Research should be conducted on the vegetation changes and pattern and on the natural zonation of mangrove communities. Knowing the regrowh stratification in areas subjecte to forest utilization, sustained extraction of desirable products could be optimized and sound management of young and old growths effected.

Studies on mariculture systems applicable in mangrove areas like fishpen and cage cultures may increase fish production with minimal interference with the ecosystem.

It is significant to study the biology and life history of commercially important fish species that spend a certain period of their early development in mangrove areas, with emphasis on migration patterns and food ecology, to rationalize the recommendations for the establishment of fish sanctuarias or reservations in specific mangrove areas. This study will also determine the role of mangrove as nurseries of commercial fishes and invertebrates.

Little is known on the effects of natural and man-induced processes on mangrove areas in the Philippines. A better understanding of the natural forces and human activities that tend to alter mangrove habitats and their biota is essential for the conservation and wise utilization of these natural resources. The research program for determining environmental effects on mangrove habitats includes biological impact of various land-uses, direct and indirect effects of human settlements, hydrodynamics, erosion and sedimentation rates, habitat alteration, pollutants and rate of land accretion on mangrove areas.

The multiple uses of the mangrove areas and their environmental impact demand sound management and disposition of these natural resources. All the other research areas in the different aspects of mangrove will serve as bases toward this end.

A major problem confronting mangrove management is the proper implementation of policies and regulations. The mangrove forests are used in highly diverse activities that a conflict ensures on the extent of utilization for each. While it is recognized that the vegetative cover should be kept intact for housing and other uses and for ecological purposes, there is also a demand for aquaculture production. Certainly, there are social and economic benefits derived from each alternative use. But there social and economic costs at the present and in the future. Socio-economic as well as ecologic/environmental values must be integrated. Management has to consider all issues and trade-offs.

Literature cited

Gabriel, B.C. 1978. Mangrove Resources: Some Issues on Their Conservation and Utilization. In Forum on Grasslands and Mangroves: Conservation and Utilization Issues, College, Laguna, Philippines.

Proceedings (various authors) 1977. Proceedings of the International Workshop on Mangrove and Estuarine Area Development for the Indo-Pacific Region, November 14–19, 1977, Manila, Philippines.

CHAPTER 12

Possible impacts of the planned hydroelectric scheme on the Purari River deltaic and coastal sea ecosystems (Papua New Guinea)

T. PETR

Office of Environment and Conservation, Central Government Waigani, Papua New Guinea[1]

Abstract. The diminished input of sediment would appear to be the major determining factor for future changes in aquatic productivity of the delta and coastal sea. If the dam were to be built, the already very dynamic geomorphology of the delta might undergo even more rapid changes than now. The currently dominant (simplified) food chain: organic matter-heterotrophic organisms-prawns-fish (crocodiles) might become second in importance with the increased transparency of water favouring primary production, and with the loss in the reservoir of much organic matter originating from the upper watershed. The contribution of tropical lowland forests, including mangroves, to the present total aquatic particular and dissolved organic matter is unknown, but if high, it would have considerable ameliorating effect on future changes in heterotrophic production. Prawn production appears to be much dependent on the present extensive areas of mangrove and tidal forests west and north-west of the Purari delta, and it is believed that any future manipulation of this until now untouched resource, would be more harmful for the commercial aquatic biota than damming the Purari.

Introduction

Among the major environmental impacts investigated in connection with plans to build a hydroelectric dam on the Purari River in Papua New Guinea are the future of the Purari mangroves and the aquatic productivity of the deltaic and coastal environments.

A dam is planned at Wabo, some 200 km upstream from the sea (Fig. 1). It would have six 360 MW generating units, five of which would be using $2\,135\,m^3/sec$. $360\,m^3/sec$. average Purari River discharge (Pickup, 1977). The rest of the water would be spilled. The effects of the dam on daily discharges in the lower Purari would be very small and probably only significant during the short periods of low flow. The volume to discharge ratio of the reservoir would be 1:5, meaning that the retention time would be only about two and a half months. This might prevent the reservoir from thermally and chemically stratifying in its major limb. Some stratification would be expected to take place in the side arms. The reservoir, which would cover an area of $290\,km^2$ at the full water supply level, would trap all coarse sediments, discharging only about 70 percent of the clay size sediments (Pickup, 1977). Before reaching the delta, the Purari would be recharged with sediments from the Aure River, which confluences with the Purari about 40 km downstream of Wabo.

The Purari delta covers approximately $1\,600\,km^2$. It consists of three major arms of the Purari, plus a

[1] Present address: Fisheries Department, (FIRI), FAO, Via delle Terme di Caracalla, 00100 Rome, Italy

H.J. Teas (ed), Physiology and management of mangroves.

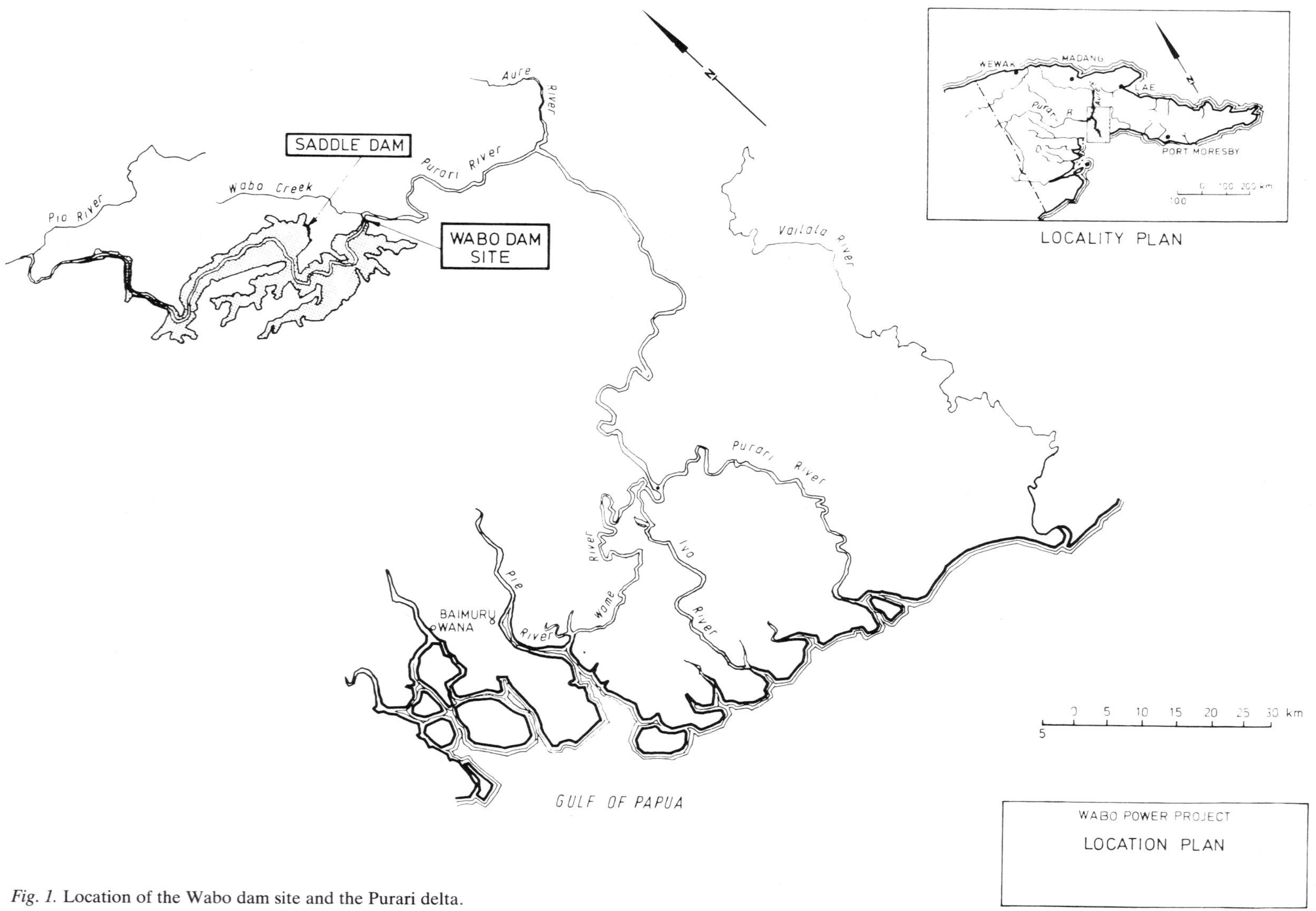

Fig. 1. Location of the Wabo dam site and the Purari delta.

number of interconnecting channels. The largest arm is the Ivo River, discharging into the Gulf of Papua much of the Purari water. The upper boundary of salinity penetration has been defined by the presence of the river bank fringing tree mangrove *Sonneratia casseolaris* and by *Pandanus* sp. (Fig. 2) (Petr and Lucero, 1979). Major concentrations of large tree mangroves such as *Bruguiera, Rhizophora, Exocoecaria, Camptostemon,* often interspersed with *Xylocarpus* and *Heritiera,* are situated in the southwestern part of the delta, especially along the tidal Pie River. Large areas of *Nypa fruticans* appear in many parts of the delta. The sago palm *(Metroxylon sagu)* is common throughout the delta, both in its freshwater and brackishwater reaches; in the latter it is found growing in up to 7.5 ppt salinity of the interstitial water (Petr and Lucero, 1979 Fig. 2). Salinity measurements by Frusher (1980) recorded a maximum inshore coastal salinity of 28 ppt near the Purari arm discharge, which also indicates that this arm carries little fresh water into the sea. Elsewhere along the coast of the Purari delta the salinity was usually less than 20 ppt.

Water discharged into the sea carries considerable quantities of fine sediments. The deltaic water transparency for light, as measured by a Secchi disc, is frequently zero, with the suspended load on some occasions exceeding 2000 ppm (Kjerfve, pers. comm.). In the Purari arm there appears to be a distinct entrapment zone, which has also been observed in the Pie tidal river bordering the Purari delta in the west. Entrapment zones elsewhere, usually of a salinity range between 1 and 6 ppt, are noted for their high concentration of nutrients, suspended solids, phytoplankton, certain zooplankton, as well as juvenile fish (Arthur and Ball, 1978). The significance of the Purari entrapment zones for aquatic productivity has not been studied yet.

The river water chemistry is dominated by calcium carbonate, the mean conductivity of the Purari water at Wabo being 125 $\mu S\,cm^{-1}$ (Petr, 1978/79). The dissolved nutrient concentrations (phosphorus, nitrogen) of the Purari are relatively high, with dissolved phosphate-phosphorus ranging from 10 to 19.2 $\mu g\,L^{-1}$, and nitrate-nitrogen from 46

to 109 $\mu g\,L^{-1}$, indicating no shortage of dissolved nutrients both for the deltaic and marine ecosystems (Viner, 1979). A high concentration of phosphorus is bound with sediments, and phosphorus metabolic availability was tested by Viner (1979) for mangrove soils bordering the delta channels and river arms. He found them fertile in terms of nutrients.

The current state of exploitation of resources of the Purari delta and the adjacent sea

In the relatively sparsely populated delta of the Purari, agricultural development remains largely at subsistence level (Stevens, 1980). The main subsistence crop is sago, which provides food, fronds for woven walls, and thatch for roofs. The chief cash crop of the delta is copra. Growing coffee, cocoa, rubber, chili, rice, ginger and balsa is largely on a trial basis. Small-scale timber cutting takes place in freshwater lowland swamps and tidal forests. A small amount of mangrove is exploited for timber, but this has hardly any impact ovethe deltaic forests. *Nypa* palm is used in small quantities as roofing material. Subsistence and commercial fisheries of the delta are underexploited, but have great potential (Stevens, 1980; Haines, 1978/79). The mangrove crab *(Scylla serrata)* has only a restricted distribution in the Purari delta, being largely captured for marketing from mangroves outside the Purari (Opnai, 1980). Some prawns are fished for subsistence within the delta, but the only large-scale commercial fishing for prawns takes place in the coastal waters of the Gulf of Papua, including those outside the Purari delta (Gwyther, 1980). Crocodiles still represent a significant source of cash income (Pernetta and Burgin, 1980).

Likely effects of damming the Purari river on the deltaic environment and the adjacent sea

In any predictions, a number of assumptions have to be made. Some of them, however, may later on prove unsubstantiated. Or, on the other hand, new unexpected situations may emerge which have not

Upper limits of *Sonneratia* and *Pandanus* sp (full line)
Upper limits of *Nypa* (dashed line)
Lowest limits of sago palm distribution in the brackish water
      environment of the Purari delta (numbers refer to sampling
      stations

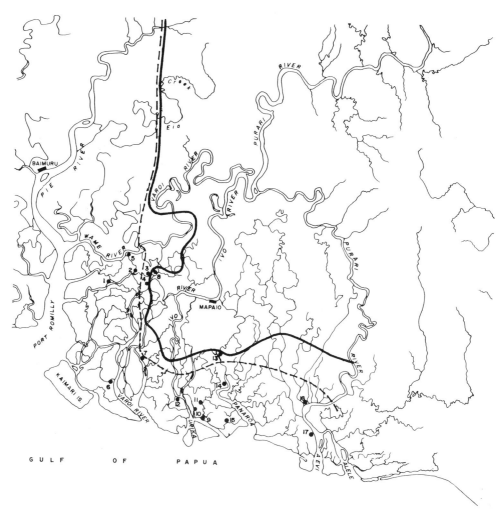

Fig. 2. Upper limits of the salinity penetration, and coastal sago palm distribution in the brackish zone of the Purari delta.

been included in the prediction.

With the increasing number of dams and reservoirs in the tropics and subtropics, prediction of environmental impact of new schemes is becoming more accurate, although much is still to be learnt. Dams in the tropics and subtropics have caused a number of environmental problems, and those connected with new lakes and the river itself have been thoroughly described in numerous papers. But only a few studies have dealt with the impact of

dams on the deltaic and coastal environments. The best known negative impact of this kind is that of the High Dam at Aswan on the collapse of the sardine catches in the Nile coastal waters (Aleem, 1969). Hall and Davis (1974) and Davies *et al.* (1975) have speculated upon the future of the extensive shrimp fishing industry in the Zambezi delta after damming the Zambezi and Kafue rivers. Davies (1979) noted in 1974 that as a result of the dams the whole sea frontage of mangroves exhibited a 1–400 m die-back. It has been speculated that the future damming of the Rufiji River in Stiegler's Gorge in Tanzania may seriously affect the mangrove forests, as well as the potential of the local prawn industry (Anon., 1980). It should be noted here that both the Zambezi and the future Rufiji dams, by storing water and therefore minimizing the downstream discharges for longer time periods, disturb the natural rhythm of seasonality, which in turn endangers the existence of floodplains with all their consequences for vegetation and animals, both terrestrial and aquatic. It also allows an upstream penetration of highly saline sea water which might be the cause of the die-back of the mangroves. In Papua New Guinea this is not to be expected as the Wabo dam on the Purari would be spilling water frequently. The Purari catchment has a very poorly pronounced seasonality which is reflected in its water discharge, with the possibility of having a peak discharge at any day of any month in a year (Government of Papua New Guinea, 1977). This excludes situations as known for the Zambezi and in future for the Rufiji deltas.

The reduction in the input of sediments to the delta would lead to channel erosion and changes in the sediment balance of the detal, leading possibly to an increase in erosion of some areas covered by mangroves. Studies on hydrology and geomorphology currently under way should provide information on the ongoing processes in this highly dynamic but still unmodified system.

At one time, fear was expressed that the damming of the Purari might increase the salinity in the delta and this could affect the sago palms growing there. As already mentioned above, sago palms in the Purari tolerate up to 7.5 ppt salinity in ground water, and it would appear that the palm does not suffer even from higher salinities during the very low Purari discharges. The regulation of the Purari River discharge by Wabo would in fact increase the absolute minimum Purari discharge rate from 430 m³/sec. (Government of Papua New Guinea, 1977) to the minimum of 2 135 m³/sec. required for running five generating units, and this would allow the sago to expand closer to the coast into or near the villages which at present have to bring their sago from higher upstream.

Major changes in the deltaic and coastal sea ecosystems of the Purari would be expected as a result of the retention in Wabo reservoir of a substantial portion of river-transported sediments. Wabo dam would reduce the present input of clay to the delta system by 22 percent, of silt by 53 percent, and of sand by 78 percent (Pickup, 1977). This would hardly affect the quantity of dissolved nutrients reaching the delta (Viner, 1979) but some reduction in particle-bound nutrients would take place. The erosion of river banks would increase gradually removing especially the finer deposited mud particles to the sea. Mud banks in brackish reaches, some of which harbour a population of commercially important post-larval stages of the banana prawn *(Penaeus merguiensis)* would gradually disappear, and with it the prawns as well. There would be an increase in water transparency, which would lead to a change in the fish community, with the 'clearer' water species gradually increasing in numbers. It is not known to what extent this would affect the fisheries.

The Purari (including the Aure) supplies a large amount of mud for the complex island system west of its delta. This area is rich in banana prawn, apparently not only because of higher water salinity which this species appears to prefer (Frusher, 1980), but also because of the prevalence there of suitable soft muddy substrata. It can be speculated that in the long run these areas might undergo erosion. But, in the meantime, probably other developments in the Purari delta would take place, such as mangrove exploitation, large-scale agricultural development, and possibly other kinds of development. If mangroves were to be removed on a scale and with the speed similar to that taking place currently in a number of other countries in the region, such an impact would probably greatly dominate over the Wabo dam impact. Martosubroto and Naamin (1977) have shown for Indonesia that shrimp production increases with the size of the tidal forests, and that 'development' of mangrove areas decreases the shrimp yield.

Mangroves provide a large quantity of organic matter which becomes colonized by bacteria and fungi.

For juveniles prawns this is the major source of protein as evidenced from experiments carried out by Edwards (1977). He kept juveniles of *Penaeus vannamei* in cages in coastal lagoons in Mexico at densities of 2.5 individuals/m². When cages were placed on the bottom, the juveniles grew 0.88 mm/day, which was similar to the growth of free prawns in the same lagoon. When the cages were lifted 30 cm from the substrate, the growth was reduced to 0.03 mm/day indicating that for fast growth juvenile prawns need a substrate rich in organic matter.

Damming the Purari would reduce the input into the deltaic system of the organic matter originating in the catchment above the dam, but it would have little effect on the input from the areas below dam, including the Aure River catchment and the delta with its tropical forests, including mangroves. The damming, in the long run, would, however, result in the gradual removal of accumulated soft clay-silt substrate (presumably rich in organic matter) which is currently favoured by the juveniles of *P. merguiensis*. In the Purari delta such areas are rather small, the major constraint there being the large input of fresh water keeping the salinity down and therefore *P. merguiensis* out. Major areas of distribution for this prawn lie outside the Purari delta, in the complex system of tidal rivers and channels spreading between the Purari and Kikori rivers (Frusher, 1980). Those areas would probably be to some degree affected by retention of sediments in the Wabo reservoir. How long it would take before any effects would take place, is difficult to say, but studies on currents and sediment transport from the Purari into this system, currently under way, may shed more light on this problem.

The fish species of the delta and the narrow belt of muddy coastal sea are those adapted to a turbid environment. Some changes in this environment are almost certain to take place due to the reduced input of sediments and clarer water after the dam is built. This might also affect the coastal sea fish, as the eastern Gulf of Papua fish species preferring clearer waters might move over into the gradually clearing waters facing the Purari. In terms of commercially important species this would mean that trevallies and tunas might become dominant over

threadfin salmon and sciaenids (Haines, 1979).

Discussion

The deltaic environment of the Purari, with its lowland freshwater swamp forest and the tidal forest, is still a largely unmodified ecosystem, one of the very few of this extent on the earth. The Purari catchment of 33 000 km² has a relatively densely populated highlands, with consequent deforestation leading to erosion and increased sediment load to the river, but the lower altitudes starting from about 1 300 m above sea level are completely covered in a tropical humid forest and are very sparsely populated.

Environmental investigations which have been carried out in connection with the proposal to dam the Purari River, have implicated the scheme in a possible impact on the Purari deltaic aquatic productivity. At present it is difficult to evaluate the magnitude of such changes, but in terms of their economic importance they would be small, much smaller than impacts of other developments, such as large-scale mangrove deforestation of the area. The damming of the Purari would diminish the input of sediment and catchment organic matter into the delta and the sea, probably leading to qualitative and quantitative changes in the present dominant food chain: inorganic/organic matter-heterotrophic production-prawns-fish (crocodiles). An increase in primary production, currently depressed by the high water turbidity, would follow, utilizing dissolved nutrients continuing to reach the delta. Heterotrophic production would somewhat decrease due to some reduction in organic matter/organic carbon inputs, as well as to the gradual erosion of bottom fine sediments, presently harbouring high concentrations of bacteria (Paerl and Kellar, 1980). Transport of organic matter and carbon by the Purari has not been studied, but concentrations of particulate organic carbon transported by the upper Amazon, which chemically appears to be similar to the Purari (Petr, 1976), were 15 to 20 g m⁻³ during the rising water level (Richey *et al.*, 1980); of this, three percent was utilized and turned over per day, which, over the long distances the Amazon traverses on its course, amounts to a large quantity of organic matter being mineralized per day. The Purari River takes less than two days to reach the sea from Wabo, and therefore the losses of particulate organic carbon are probably minimal, and much of the organic matter which enters this short and fast mountainous river from its watershed probably reaches the delta and sea. A reservoir at Wabo would retain much of this organic particulate matter. Below the dam, a gradual removal of old sediments by erosion, and oxidation of the old organic matter would take place, leading to changes in benthic substrate, which, in turn would lead to changes in benthic fauna, both perennial and temporary. The microbiota of fine sediments would also start undergoing changes due to the physico-chemical alteration of the substrate, thus facilitating disruption of the present food chain. The fish, apart from responding to the change in their food base, would also react directly to the physical and chemical changes, especially to the gradually increasing transparency of water. The result would be not only a new fish community – probably based on species of clearer waters, but also a shift in the ratio between filter feeders and predators.

The river plume would become smaller due to the entrapment of most Purari sediments in the reservoir. With the discharge of water into the sea largely still unregulated by the dam, the dissolved nutrients might be reaching the sea at about the same concentration as now, subject to the rate of primary production and the physico-chemical properties of the water mass in the reservoir. As a result of increased transparency the depth of the euphotic zone would increase both in the delta and in the plume, resulting in higher primary production. But reduced amounts of suspended inorganic and organic particulate material would result in reduced heterotrophic production of the plume and in gradual oxidation of organic material present in bottom deposits there. This would probably be followed by the gradually diminishing numbers of prawns and consequently by lower commercial catches. The time scale for these changes is difficult to even guess. The bottom substrate would probably loose its organic matter slowly, over many years. An important ameliorating factor would be the continuity in the input of the Aure River sediments, which contribute substantially to the Purari below Wabo (Pickup, 1977).

The organic matter input from mangroves, the rate of which is currently unknown, would ameliorate the losses from the upper Purari watershed. There is also a possibility of appearance of sea grasses on substrata overlaid by clearer waters. Sea grasses would than become a new substrate for prawns, as well as a source of organic matter and carbon.

The contribution of the entrapment zone of the very complex Purari-Kikori deltaic system to the brackishwater aquatic production, is at present unknown. It is possible that the entrapment zone functions to some degree as a self-contained system, which would continue utilizing the nutrients of soft bottom muds of this zone to maintain a fairly high productivity.

The transport of organic matter, the carbon input from the watershed, and the role of mangroves as a largely independent source of organic matter, are among the many topics still to be investigated. The physico-chemical and biological environments of the Purari plume and of the turbid zone of the extensive Purari-Kikori deltaic system are virtually unknown, although these are of much importance for the existence of commercial stocks of prawns, fish and crabs. Studies in progress on deltaic geomorphology and hydrology will greatly assist in the future modelling of the potential impact of the dam on the coastal systems and it would be desirable to improve the quality of the model by generating more inputs from additional in-depth studies. As the project is still in the planning stage, this gives the advantage of having time for further studies.

Acknowledgments

This paper is based on investigations of a large team of researchers, each member of which contributed in some way. Their contributions, and the financial support of the Government of Papua New Guinea which made the work possible, are greatly appreciated.

Literature cited

Aleem, A.A. 1969. Marine resources of the United Arab Republic. Stud. Rev. Gen. Fish. Coun. Mediterr. 43:1–22.

Anon. 1980. Tanzania project to weigh impact on environment. Blueprint for better planning? Ceres, March-April. p. 12.

Arthur, J.F. and M.D. Ball. 1978. Entrapment of suspended materials in the San Francisco Bay-Delta Estuary. U.S. Dept. Interior, Bureau of Reclamation. 106 pp.

Davies, B.R. 1979. Stream regulation in Africa: A review. In The ecology of regulated streams (eds. J.W. Ward and J.A. Stanford). Plenum Press, New York and London. pp. 113–142.

Davies, B.R., A. Hall and P.B.N. Jackson. 1975. Some ecological aspects of Cabora Bass Dam. Biol. Conserv. 8:189–201.

Edwards, R.R.C. 1977. Field experiments on growth and mortality of *Penaeus vannamei* in a Mexican coastal lagoon complex. Est. Coast. Mar. Sci. 5:107–121.

Frusher, S.D. 1980. The inshore prawn resource and its relation to the Purari delta region. In Possible effects of the Purari Hydroelectric Scheme on subsistence and commercial crustacean fisheries in the Gulf of Papua: Workshop 12 December 1979 (ed. D. Gwyther). Purari River (Wabo) Hydroelectric Scheme environmental Studies Vol. 15. Office of Environment and Conservation, Waigani, and Department of Minerals and Energy, Konedobu, Papua New Guinea. pp. 11–27.

Government of Papua New Guinea. 1977. Purari River Wabo Power Project, Feasibility Report. Vol. 1. Summary. 10 p., 19 Figs., App. A–C.

Haines, A.K. 1978/79. The subsistence fishery of the Purari delta. Science in New Guinea 6:80–104.

Haines, A.K. 1979. An ecological survey of fish of the lower Purari River system, Papua New Guinea. Purari River (Wabo) Hydroelectric Scheme Environmental Studies Vol. 6. Office of Environment and Conservation, Waigani, and Department of Minerals and Energy, Konedobu, Papua New Guinea.

Hall, A. and B.R. Davies. 1974. Cabora Bassa: Apreciacao global do seu impacto no vale do Zambeze. Rev. Econ. Moçambique 11:15–25.

Martosubroto, P. and N. Naamin. 1977. Relationship between tidal forests (mangroves) and commercial shrimp production in Indonesia. Marine Res. in Indonesia (No.18):81–86.

Opnai, L.J. 1980. The mangrove crab *Scylla serrata*. In Possible effects of the Purari Hydroelectric Scheme on subsistence and commercial crustacean fisheries in the Gulf of Papua: Workshop 12 December 1979 (ed. D. Gwyther). Purari River (Wabo) Hydroelectric Scheme Environmental Studies Vol. 15. Office of Environment and Conservation, Waigani, and Department of Minerals and Energy, Konedobu, Papua New Guinea. pp. 83–91.

Paerl, H.W. and P.E. Kellar. 1980. Some aspects of the microbial ecology of the Purari River, Papua New Guinea. In Aquatic Ecology of the Purari River Catchment. Purari River (Wabo) Hydroelectric Scheme Environmental Studies Vol. 11. Office of Environment and Conservation, Waigani, and Department of Minerals and Energy, Konedobu, Papua New Guinea. pp. 25–39.

Pernetta, J. and S. Burgin. 1980. Census of crocodile populations and their utilisation in the Purari area. Purari River (Wabo) Hydroelectric Scheme Environmental Studies Vol. 14. Office of Environment and Conservation, Waigani, and Department of Minerals and Energy, Konedobu, Papua New Guinea.

Petr, T. 1976. Some chemical features of two Papua fresh waters (Papua New Guinea). Austr. J. Mar. Freshwat. Res. 27:467–474.

Petr, T. 1978/79. The Purari River hydroelectric development at Wabo and its environmental impact: An assessment of a scheme in planning stage. Science in New Guinea 6:105–116.

Petr, T. 1980. Purari River Environment (Papua New Guinea). A summary report of research and surveys during 1977–1979. Office of Environment and Conservation, Waigani, and Department of Minerals and Energy, Konedobu, Papua New Guinea. 57 p.

Petr, T. and J. Lucero. 1979. Sago palm salinity tolerance in the Purari River delta. In Ecology of the Purari River Catchment. Purari River (Wabo) Hydroelectric Scheme Environmental Studies Vol. 10. Office of Environment and Conservation, Waigani, and Department of Minerals and Energy, Konedobu, Papua New Guinea. pp. 101–106.

Pickup, G. 1977. Computer simulation of the impact of the Wabo hydroelectric scheme on the sediment balance of the Lower Purari. Purari River (Wabo) Hydroelectric Scheme Environmental Studies Vol. 2. Office of Environment and Conservation, Waigani, and Department of Minerals and Energy, Konedobu, Papua New Guinea.

Richey, J.E., J.I. Brock, R.J. Naiman, R.C. Wissmar and R.F. Stallard. 1980. Organic carbon oxidation and transport in the Amazon River. Science 207:1348–1351.

Stevens, R.N. 1980. The agriculture and fishery development in the Purari delta in 1978–1979. Purari River (Wabo) Hydroelectric Scheme Environmental Studies Vol. 13. Office of Environment and Conservation, Waigani, and Department of Minerals and Energy, Konedobu, Papua New Guinea.

Viner, A.B. 1979. The status and transport of nutrients through the Purari River. Purari River (Wabo) Hydroelectric Scheme Environmental Studies Vol. 9. Office of Environment and Conservation, Waigani, and Department of Minerals and Energy, Konedobu, Papua New Guinea.

CHAPTER 13

Overcoming problems in the management of New Zealand mangrove forests*

P.R. DINGWALL

Department of Lands & Survey Wellington, New Zealand

Introduction

Since the first international symposium on the biology and management of mangroves, held in Hawaii in 1974, there has been a far greater awareness of the ecological role and significance of mangroves, and the urgent need for the application of sound management practices to ensure that mangrove environments are not depleted or destroyed.

At least two subsequent international forums have added their voice to the call to world governments for the wise use and sustainable development of mangrove forests. The 12th General Assembly of the International Union for the Conservation of Nature and Natural Resources (IUCN), meeting in Zaire in 1975, drew attention to the need for control of destructive practices; the desirability of surveys and evaluations of mangrove resources; and the urgent requirements to establish and manage large ecologically viable reserves over mangrove and associated communities. The UNESCO Regional Seminar on Human Uses of Mangrove Environment and its Management Implications, held in Dacca in December 1978, recommended the establishment of national mangrove committees which would be responsible for compiling bibliographies of relevant research; facilitating the collection and analysis of data on mangrove environments; designing biosphere re-

serves in mangrove areas; and coordinating research and management programmes. The Seminar also stressed that governments should promote education and training in matters of mangrove management, and recommended that management plans be drawn up for mangrove areas to guide the use of mangrove forests and forecast the likely long-term costs resulting from changes induced by man.

This paper provides a critical review of recent developments in the management of New Zealand's mangrove forests, in the light of these international policies and objectives.

Characteristics of the New Zealand mangroves

The native New Zealand mangrove, and the only one occurring, is the black mangrove *Avicennia*, referred to usually as *A. resinifera* Forst. f. but also known as *A. marina* (Forsk.) Vierh. var. *resinifera* (Forst. f.) Bakh. (Allan 1961). Thus, its exact taxonomic status and affinity with the neighbouring Australian species *A. marina* remain as unanswered and equivocal questions.

The New Zealand mangrove is a broadleaf, evergreen tree. Specially adapted to living in saline, anaerobic conditions, it is usually the pioneering plant on unconsolidated mudbanks, and it forms

---

* (Paper prepared for Second International Symposium on biology, and management of mangroves and tropical shallow water communities, Port Moresby, Papua-New Guinea, July 1980)

H.J. Teas (ed), Physiology and management of mangroves.
© *1984 Dr W. Junk Publishers, The Hague. ISBN 90 6193 949 6. Printed in the Netherlands.*

the seaward belt in a plant succession which culminates in the dry land of salt marshes.

North of Whangarei it is a considerable tree, reaching a maximum height of around 10 m. Many observers refer to a steady decrease in the height of mangroves to about 0.5 m at their southern limit. However, this is true only in the very broadest sense, as there is a marked variability in the stature of trees throughout their range. Clearly, factors other than climate are also important in determining the height and vigour of mangroves. They tend to be larger along their outer margins and on the borders of streams, where drainage is favourable. A close positive relationship between the height of trees and thickness of soft sediments has also been reported (pers. comm. N.H. Nickerson, Tufts University, Massachusetts, USA). A feature of New Zealand mangroves is their tendency toward dwarfism, and there are broad expanses of mangrove forest with trees generally less than 1 m tall. The reasons for this are still largely unknown, but there are several possible explanations including, a younger age, the presence of coarser sediments or a thin substrate, more sluggish drainage, a small sea level rise allowing expansion over the shore, or even the impact of cattle grazing.

Mangrove ecology

The ecology of New Zealand mangrove forests has been comprehensively discussed elsewhere (Chapman and Ronaldson 1958; Morton and Miller 1968; Kuchler 1972; Ritchie 1976) and will be treated only briefly here.

Kuchler (1972) has made a useful distinction between the two major groups of mangrove communities in New Zealand – the mangrove swamp and the saltmarsh community. The mangrove swamp occupies the shore zone experiencing regular tidal flooding, where mangroves are associated with a variety of algae. Saltmarsh communities are ubiquitous throughout the mangrove range but assume greater importance farther south. They are less frequently flooded, and the mangrove, though conspicuous, may be dominated by a complex of rushes, sedges, succulents, herbs and grasses including

Salicornia australis, *Juncus maritimus* var *australiensis*, *Scirpus cernuus*, *Leptocarpus similis* and *Samolus repens*. The term 'mangrove savanna' has been applied to situations where mangroves occur as scattered, isolated trees in saltmarsh.

Mangrove swamps are rich in invertebrates, including crustaceans such as the mud crab (*Helice crassa*) and the snapping shrimp (*Alpheus sp.*); and molluscs such as the mud whelk (*Cominella glandiformis*), the spire snail (*Zeacumantus lutulentis*), the mudflat topshell (*Zediloma subrostrata*) and the common cockle (*Chione stutchburyi*). Ritchie (1976) notes that at least 30 species of fish utilise mangrove swamps either as permanent residents (*e.g.* flounder, mullett, eels), as frequent visitors (*e.g.* snapper, trevally, baracouta, mackerel), as occasional visitors (*e.g.* dogfish, shark, red moki) or as migratory fish (*e.g.* freshwater eels, galaxids and bullies). Hosts of birds frequent mangrove forests, including forest birds (*e.g.* fantail), waders (*e.g.* oystercatcher, pied stilt) and others such as gulls and shags.

Distribution of mangrove forests

Mangroves are widespread along the shores of sheltered coastal waters in northern New Zealand (Fig. 1). Their distribution is unknown in detail as they have not been mapped. Nor has their real extent been precisely calculated, though it probably totals approximately 20,000 ha. Mangroves occur in virtually every harbour and estuary in the North and South Auckland regions, attaining their greatest extent in Kaipara, Hokianga and Rangaunu Harbours.

The southern limit of their distribution is a matter of some interest as it approaches the extreme poleward extent of mangroves in the southern hemisphere. On the west coast, mangroves have commonly been reported as far south as Kawhia Harbour but a recent survey failed to locate them there. This means that the southernmost naturally-occurring community of mangroves is now at Raglan Harbour, at 37° 40′ S. However, a single plant is known from the Awakino River at 38° 39′ S, and there have been recent plantings of

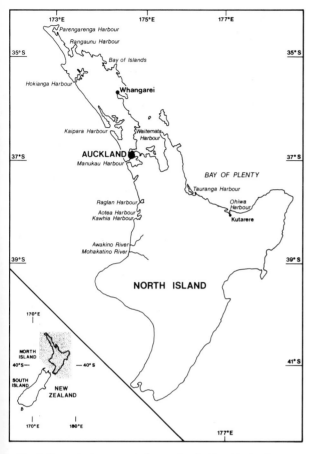

Fig. 1. Important mangrove forest sites in New Zealand.

(Wells 1980). On the east cost of New Zealand, mangroves extend south to Kutarere on Ohiwa Harbour, at 38° 03′ S.

While there are undoubtedly several controls of mangrove distribution, including the pattern of coastal water flow and the availability of favourable sheltered water and fine sediments, it is air temperature which appears to be the ultimate control of their southernmost limit. Mangroves are unable to withstand prolonged periods of freezing. *Avicennia* is able to tolerate occasional frosts, however. The climate throughout its range is warm temperate rather than sub-tropical, with warm to hot summers and mild winters. The mean temperature of the coldest month ie July, is around 9–11° C and frosts, though uncommon, do occur (Table 1). Although there is no strong evidence to suggest that mangroves were formerly of much greater extent in New Zealand, it does appear that they are undergoing some recession along their southern margin, as evidenced by the disappearance of stands in Kawhia Harbour during the 1970's. In the absence of any significant shift in the climatic regime the stresses being experienced here are probably man-induced, with accelerated sedimentation as the probable, though unproven, cause.

mangroves for streambank protection in the Mohakatino River at 38° 43′ S. These are among the southernmost mangroves in the world. The southernmost community of *A. marina* in Australia is in Corner Inlet, Wilson's Promontory at 38° 45′ S

The character of mangrove environments

In that part of New Zealand where climate allows mangrove growth, the topography of the coast, replete with harbours, estuaries and embayments,

Table 1. Mean annual and mean July air temperatures, and frequency of frosts at some important mangrove sites in New Zealand. (Source: NZ Meteorological Service 1973).

Location	Temperature (°C)		Days with Frost	
	Mean Annual	Mean July	Air	Ground
Parengarenga Harbour	15.2	11.5	2	5
Bay of Islands	14.7	10.7	1	26
Kaipara Harbour	14.3	10.3	3	9
Manukau Harbour	14.6	10.2	2	22
Raglan Harbour	14.3	10.1	3	12
Tauranga Harbour	14.0	9.2	7	60
Ohiwa Harbour	14.1	9.2	6	–

provides much sheltered water for mangroves to establish.

The North Auckland Peninsula exhibits markedly constrasting configuration on its eastern and western coasts. On the west coast, post-glacial submergence of deeply dissected terrain has formed several extensive harbours, widely separated by long, straight stretches of coast where the bedrock is covered by a broad zone of consolidated Pleistocene sands and Holocene dunes. Much of the coast here is too exposed and sandy for mangrove establishment and mangroves are confined to the harbours where they have colonised extensive areas of mudflats. The large Raglan, Aotea and Kawhia Harbours further south are also 'drowned' lower reaches of river valleys cuts during glacial times of lower sea levels.

The east coast is steeply cliffed and intricately embayed. Unlike the western harbours, infilling of these bays has generally occurred only at the extreme bayhead, reflecting the lower sediment contribution from the smaller eastward-draining catchments. Mangroves have colonised these flats throughout, but are better developed where the substrate is muddy than where sand dominates; such as at Parengarenga, where there are broad sandy expanses without mangroves. The Bay of Plenty coastline is formed from the dissected margin of an ignimbrite plateau which is inclined seaward and is 'drowned' by post-glacial transgression. Mangroves here are confined to estuaries which provide sheltered sites behind spits and bars.

Tides in northern New Zealand are semi-diurnal, with a mean spring tidal range of approximately 2 m, though varying from 1.6 m at the Bay of Plenty to 2.9 m in Raglan Harbour (Marine Division, Ministry of Transport, 1979). The extreme spring tide range is as much as 3.6 m in Hokianga and Manukau Harbours (Morton and Miller 1968). Since mangrove growth reaches a lower limit at about half-tide, there is normally a vertical range of 1.5 m in the forest communities.

Nearshore currents flow north along the west coast, though in the extreme north a southerly flow, the West Auckland current, may occur in winter (Heath 1971). The general northward drift

here is unfavourable for the southward spread of mangroves. On the east coast, however, the predominant coastal current, the East Auckland current, flows south and into the Bay of Plenty. Surface water temperatures range from approximately 20° in summer to 15° in winter (Garner 1969).

The values of mangrove forests

Morton (1976) has comprehensively summarised the widely ranging values of mangrove forests, as follows.

(i) Ecological value
Mangroves constitute a highly productive estuarine ecosystem of vital importance in animal food webs. Their biological richness is not generally appreciated. Ritchie (1976) notes that New Zealand mangroves produce some 10 tonnes of organic plant debris/ha/year and enrich the nearshore environment by a factor of at least 10 over similar coastlines lacking mangroves. They are also the breathing areas of harbours, re-oxygenating waters during the ebb and flood of tides.

(ii) Sedimentation and coastal protection value
With their expansive, shallow root systems festooned with pneumatophores, mangroves have the capacity to trap fine sediment and so build out coastal areas. They thus form an effective buffer zone protecting the shore from the erosive forces of wave action. Indeed, mangroves are often purposefully planted in shore protection works or to shelter wharves and marinas.

(iii) Wildlife habitat value
Mangrove forests are refuges for a host of bird species including forest birds in the canopy, and waders in the tidal muds. In addition to local species, global migrants, such as godwits and knots, also inhabit mangrove areas.

(iv) Economic value
There is essentially no direct economic use of mangroves in New Zealand. With several alternative sources readily available, there is no demand for

the wood of mangroves for fuel, pulpwood, or building timber. According to Chapman (1976) and Walsh (1977) *Avicennia* is not a preferred fuel, nor is it suitable for pulping. Elsewhere, especially in developing countries, it has a few specialised uses, in shipbuilding for example, and for various medicinal purposes. *A. germinans* is reported to have value as a producer of high quality honey. However, Walsh (1967) notes that honey from the New Zealand mangrove has a most unpalatable flavour and has no commercial prospects.

(v) Coastal fisheries value

Much coastal fishery in New Zealand is fundamentally dependent upon mangrove forests. According to Ritchie (1976), at least 30 species of fish use mangrove wetlands at some stage in their life cycle. In North Auckland some commercially valuable species such as eels, flatfish, grey mullet, pipi and cockle are caught almost exclusively in mangrove areas. Important catches of kahawai, parore, snapper and trevally are also made. It is also probably, but unsubstantiated by research, that mangroves make an important contribution to large fisheries of school shark, mussels and pipis at the entrances to several large northern harbours. In the Arapaoa River, Kaipara Harbour, oyster farms and flounder fishing support 12 full-time family units (and many more in part) and yield $100,000 annually (Ritchie 1976).

(vi) Aesthetic value

Mangrove forests have traditionally been regarded as unpleasant wastelands. There is even evidence to suggest that clearing of mangroves was often an unwarranted fearful response to the supposed health hazard from mosquitoes and other inhabitants of mangrove wetlands. Despite this long tradition of disdain and fear, many people now view the mangrove forest as a scenically attractive one, making a pleasing panoramic contribution to the shorescape. Indeed, in northern New Zealand the mangrove is an integral component of the natural history and landscape of the coast. It is not only the trees which attract, but also the other associated vegetation and the birds which add to the total aesthetic qualities of these areas.

(vii) Scientific and educational values

As a unique amphibious intertidal forest, mangroves are of great botanical interest. They display a fascinating array of physiological and structural adaptions to their demanding and specialised environment, including peculiar breathing roots, or pneumatophores, underground aeration tissues for conducting and storing oxygen, resistant corky bark, and large viviparous seedlings which promote effective dispersal. There is also much scientific interest in the inter-relationships between plants and animals, vegetation dynamics and plant succession, and the role of geomorphic site and history in the establishment and development of mangrove forests. But, above all, the New Zealand mangroves are scientifically important because here *Avicennia* is at the poleward limit of its global geographic range, ie at its environmental extreme limits and possibly specially adapted to a temperate climate.

Human impacts on mangroves

Man can affect mangrove forests either directly by reclaiming the land and replacing mangroves with his own structures, or indirectly through modifying the waters that drain through mangroves either from the land or sea. The more conspicuous impacts, certainly in New Zealand, are those associated with reclamation.

There is a long history of mangrove reclamation in New Zealand, but it is poorly documented and therefore very difficult to assess objectively. Morton (1976) notes that mangroves are a shrinking community and that much of it is already lost, but he provides no evidence to substantiate this claim. In a rare attempt to qualify the impact of reclamation, the Nature Conservation Council (1976) reported that in the Hokianga Harbour, of the original 20.5 square miles of mangrove and saltmarsh only 10.64 square miles remain. Chapman and Ronaldson (1958) identify several areas in the Waitemata Harbour modified by the expansion of the city of Auckland. Thus, while it is clear that reclamation has taken a heavy localised toll on mangroves, it is difficult to obtain an overall measure of its impact.

The principal reason for reclaming mangroves in New Zealand was to replace them with pasture. In this sense, mangrove replacement was but part of a broader process of conversion of indigenous forests. To the agricultural pioneer, forest land was wasteland requiring conversion to more productive pasture lands. Some two-thirds of New Zealand's native forest cover have been converted for pastoral farming purposes.

The reclamation of intertidal mangrove forests in North Auckland was especially attractive as it was a means of adding easily ploughable flat land to predominantly steep hill country farms. In the Hokianga area, agricultural reclamation of mangroves began during the depression of the early 1920's (Sherwood 1963). The process involved construction, by hand, of outer retaining walls; internal drainage to promote the leaching out of salt from the soils; and finally, after 3 to 10 years had elapsed, the planting of grasses and clovers. It was soon realised that high quality swards could be grown (Hamblyn 1932), resulting in a huge boost to farm production and easier farm management.

Agricultural reclamation of mangroves has been essentially phased out now, but there are still a few properties in the process of conversion. To one such property at the head of Rangaunu Harbour some 320 ha have been added over the past 25 years. Experience has shown that with careful attention to drainage, and heavy applications of lime and superphosphate, pastures can become productive within seven years of commencing reclamation.

Even without reclamation it is possible for farming activities to affect mangroves. Much of the intertidal zone is unfenced and cattle straying from adjoining farms have grazed mangroves over broad areas, notably in Parengarenga Harbour. The salt content in mangroves appears to be a tonic as the cattle often become addicted to eating them. It is not unusual for cattle to be drowned when trapped on mudflats by a rising tide. Grazed mangroves are usually less than 1 m tall and exhibit a characteristic prostrate form where the nipping off of top buds and branches induces lateral spreading.

While pastoral farming is the principal reason for reclaiming New Zealand mangrove forests, it is not the only one. Much reclamation is undertaken in association with the construction of causeways across the heads of tidal inlets, for road and rail traffic. Normally, in these instances the opportunity is taken to reclaim the mangrove stand which is thus isolated from the sea. Especially in the vicinity of urban areas, reclamation for housing and industrial sites, oxidation ponds, or for marinas is common. The intertidal zone is also a favoured dumping ground for domestic and industrial wastes and rubbish tips abound in mangrove forests. Not only are mangroves buried in this process, but the leaching out of pollutants often destroys adjacent mangroves and their associated faunas.

Man's indirect impact on mangroves in New Zealand is extremely difficult to assess in the absence of any complete records of the former distribution of mangrove forests.

Since mangroves are highly specialised plants with very specific habitat requirements, they are very sensitive to any environmental changes. Perhaps the most significant change affecting the mangrove environment has been an increase in sedimentation rates consequent upon the settlement of land. During the early and middle decades of this century, the clearing of the natural forest cover over whole catchments and conversion to farmland resulted in accelerated soil erosion and associated excessive downstream sedimentation. The slow-growing mangroves were often unable to keep pace with the rapid accumulation of terrestrial silts and clays and the burial of their root and pneumatophore systems caused them to die off. This process is well-illustrated in the lower reaches of the Puhoi River, north of Auckland. The implementation of improved soil conservation techniques and land development practices, and the spread of a new grass cover in recent times has slowed down this process but its residual effects will undoubtedly continue for some time. To these will be added a new threat, of unpredictable proportions, from chemical runoff derived from artificially topdressed pastures.

Mangrove reserves

At present there are only two reserved areas of mangroves in New Zealand, that are of any significance. Together they contain only about 550 ha of mangrove forest, or some 2.5% of the total area of mangrove in the country. This can hardly be regarded as a truly representative sample.

The first such reserve is at Tauhoa in the Kaipara Harbour, set apart in 1948 as a Reserve for the Preservation of Native Flora, under Section 360 of the Land Act 1924. The reserve, which contains both mangrove and saltmarsh communities, covers a total area of 301.5 ha of which all but 10 ha is within the intertidal zone. The land was secured following the initiative of Professor V.J. Chapman of Auckland University, and control of the reserve was subsequently vested in the Council of the University. Unfortunately, little use has been made of the reserve for ecological research or monitoring, and in the early 1970's, in response to applications for reclamation, the University had even agreed in principle to the conversion of a substantial portion of the reserve to farmland. Fortunately, this reclamation never eventuated but a temporary Licence to Occupy was issued to one landowner in 1972, for grazing purposes. This reserve is now an extremely valuable and secure one. For, with the carrying forward of provisions from the original Act to subsequent Acts, the reserve is now a Nature Reserve under the Reserves Act 1977. This classification is applied only to the most ecologically important and internationally significant reserves in the country and the legislation provides for strict protection of the area in its natural state.

The second important reserve with mangroves is the Waitangi National Reserve in the Bay of Islands. This reserve is primarily of historic significance as it is the site of formal annexation of New Zealand to the British Colonial Empire. However, the reserve also contains some 245 ha of intertidal area, most of which is covered in mature mangrove forest. A feature of this reserve is a magnificent boardwalk, 280 m long, established to provide public access into the mangrove forest and as a basis for an educational and interpretative programme.

Two other areas of mangrove forest, both in Ohiwa Harbour, are under negotiation as new reserve areas. These are important areas since mangroves here are at the southern extremity of their distribution on the east coast. The first is Motuotu Island, at present a wildlife refuge, which together with its surrounding foreshore is proposed as a Nature Reserve. Mangroves cover some 53 ha of this area. The second area is a proposed Scientific Reserve, which will involve reservation of 10 ha of mangroves and their addition to the present Patawa Island Scenic Reserve. This will be an invaluable site for scientific research and monitoring of the mangrove ecosystem.

These areas, though important, will make only minor additions to the reserve network. In recent years the New Zealand Nature Conservation Council, concerned over the disproportionately small areas of mangroves in reserves, has urged that the government undertake further reserve surveys. The Department of Lands and Survey responded by commissioning Professor V.J. Chapman to survey all the important northern harbours and to make recommendations for reserves. This represents the first attempt to comprehensively survey the New Zealand mangrove forests, and his recommendations warrant close attention.

Legislation, planning and management

While there is an unquestioned need to secure more areas of mangrove forests in reserves, this is not a simple legislative or administrative matter in New Zealand.

Management of the foreshore, which is public land under Crown ownership, has become an administrative nightmare. The principal cause of this is a legislative framework designed to regulate the use of resources rather than areas, which results in fragmented control over any one part of the intertidal zone. These are at least fifteen statutes for the control of activities in tidal areas. Thus, for example, the undertaking of physical works and the use of boats are controlled under the Harbours Act 1950; the taking of fish is regulated under both the Fisheries Act 1908 and the Marine Farming Act 1971; waste discharges are regulated under the water and Soil Conservation Act 1967; and coastal protection is controlled under the Soil Conservation and Rivers Control Act 1941.

Since 1970, recognition of the urgent need to streamline and coordinate management in the coastal zone has resulted in some encouraging initiatives in maritime planning. Responsibility for maritime planning is now vested jointly in the Ministry of Transport and the Minister of Works and Development, under Part V of the Town and Country Planning Act 1977. Although it is still too early to judge the effectiveness of this new arrangement, there now exists in New Zealand formal planning procedures for maritime areas, akin to the longer-established District Scheme planning on land.

Several of the statutes controlling foreshore areas have a direct bearing on the management and conservation of mangrove forests. That having the greatest influence is the Harbours Act 1950, administered by the Ministry of Transport, which provides for approval of reclamations, impoundments and the erection of structures. A fundamental change in approach toward such approvals occurred in 1973 when an amendment to the principal act made it obligatory for the 'public interest' to be taken account of in considering applications for construction and other permanent uses of the foreshore or seabed.

Procedures for approving reclamations illustrate this new approach well (Ministry of Transport 1978). The Act prohibits the filling in of any tidal waterway without proper authority. There are five different ways in which reclamations can be authorised:

(i)   Areas larger than 4 ha are normally authorised by a Local Act of Parliament;

(ii)   Areas less than 4 ha to be reclaimed 'for the benefit of the public' are authorised by Order-In-Council without advertising for objections. This is a rarely used procedure;

(iii)   Areas less than 4 ha are authorised by Order-In-Council after objections are called for. This is the most common procedure used;

(iv)   Harbour Boards may be authorised by Order-In-Council to carry out urgent reclamations over 4 ha, but only after objections are heard;

(v)   Very small reclamations incorporated in the building of a structure may be authorised as part of the approval of plans for the structure.

The criteria used in evaluating applications for reclamations are as follows:

(i)   The proposed reclamation must be as small as possible and fully justified with no available alternative;

(ii)   No reclamations of public space will be approved for private gain. Thus housing developments are generally opposed;

(iii)   Applications for refuse tips are opposed, and this activity is to be progressively phased out of the tidal zone;

(iv)   Applications for the disposal of spoil are normally opposed;

(v)   Causeways, where approved, must allow for the normal flow of tidal waters to cut-off areas.

Firm applications for reclamations must be accompanied by information on the likely environmental impacts and at an early stage the Ministry seeks guidance from the Commission for the Environment over the desirability of full environmental impact reporting procedures.

The Act makes it an offence for any person to undertake reclamation works or build a structure without proper authority and provides for substantial fines in the event of a prosecution.

Thus, the official attitude toward the foreshore is that, as publicly owned space it is equivalent in status to a reserve on land and any major modification or change of use warrants close scrutiny.

However, the Harbours Act does not provide for the setting apart of reserves *per se*. There is a strong argument for extending a protective status over discrete areas of the intertidal realm and the application of management policies and programmes which guide the use of terrestrial reserves.

The principal act for setting apart terrestrial reserves in New Zealand is the Reserves Act 1977, which provides for seven classes of reserve varying according to their principal purpose and the degree of security accorded to their preservation in a natural state. There is no authority to set apart areas below high water under this Act, but much could be gained from it. For example, a reserved area of foreshore would benefit from day-to-day supervision by rangers, who would have a responsibility for implementing a comprehensive management plan drawn up in consultation with the public. Areas of foreshore would also become an integral part of a national network of reserves and benefit from a holistic approach to reserve use, research, information gathering and dissemination, public education etc. Moreover, the high tide mark which has become something of an administrative barrier, is not an ecological one. Effective management of terrestrial coastal reserves is often dependent upon compatible and integrated management of adjacent waters.

The Ministry of Transport is now giving emphasis to the coordination of adjacent land and water activities and there is now provision under the Harbours Act for control of foreshores and coastal waters to be delegated to local authorities or park and reserve boards. Negotiations are now underway for appropriate areas of foreshores to be set apart under Section 167 of the Land Act 1948 for reserve purposes and then to be managed under the Reserves Act, and this holds much promise for the seaward extension of land reserves.

A statute with real, but unrealised potential for reserving foreshore areas is the Marine Reserves Act 1971. According to the long title of this Act, it provides for the setting up and management of areas of the sea and foreshore as marine reserves for the purpose of preserving them in their natural state as the habitat of marine life for scientific study. Thus, it essentially allows for only Scientific Reserves. It was designed to reserve a particular area for scientific research and has proved far too restrictive to be workable. The Act was amended in 1977 to allow fishing in a marine reserve, but it is still not possible to establish a marine reserve where the principal objective is recreation, tourism, or nature conservation *per se*. The Act should either be rewritten to provide for the diversity of uses of reserves allowed under the Reserves Act 1977, or abolished.

In an attempt to partially fill the gap in the conservation of marine resources, the Fisheries Amendment Act 1979 extended the definition of 'fish' to include 'every description of flora and fauna naturally occuring seawards of mean high water'. Thus in the legal sense a mangrove is now a fish and subject to regulations of the Fisheries Act! It is unrealistic, however to assume that this Act will be effective in the management and protection of mangrove forests. Nor can we expect that such provisions will substitute for the real need – an adequate system of marine reserves complementing the network of national parks and reserves on land.

Conclusion

The discussion above has revealed that in recent years there has been a considerable improvement in the management of New Zealand mangrove forests and that real progress has been made in responding to international calls for a controlled and beneficial use of mangrove ecosystems.

Mangroves have become an outstanding, albeit controversial, environmental topic. Past management was often wasteful and destructive and based largely on the view that mangroves were an expendable resource of little value. A marked changed in attitude has occurred recently, reflecting in part a generally increased sophistication in the use of natural areas and resources, but also acknowledging a greater awareness of the vital function and intrinsic values of the mangrove forests themselves.

This change in attitude is now becoming reflected in the more enlightened legislation, planning procedures, administrative organisation and management systems.

There is still no room for complacency, however,

as much further progress is required. More effort must be made in research aimed at increasing our understanding of the role of mangroves in the coastal realm and their significance for local economic and social well-being. There is also a need to bring greater areas of mangroves within the existing network of parks and reserves and this requires solutions for present administrative problems opposing this objective.

Despite past ignorance and wastefulness, much of the mangrove resource remains intact and there is an opportunity for New Zealand to continue its contribution to international efforts at improving mangrove management. The decade of the 1980's will be a testing one in New Zealand, but recent experience in mangrove management gives some reason for confidence in a successful outcome.

Literature cited

Allan, H.H. 1961. *Flora of New Zealand,* Vol. 1, Government Printer, Wellington. 1085 pp.

Chapman, V.J. 1976. *Mangrove Vegetation,* J. Cramer, Vaduz, 447 pp.

Chapman, V.J. and J.W. Ronaldson. 1958. The mangroves and saltmarsh flats of the Auckland ithsmus. New Zealand Department of Scientific and Industrial Research, Bulletin 125; 79 pp.

Garner, D.M. 1969. The seasonal range of sea temperature on the New Zealand shelf. New Zealand Journal of Marine and Freshwater Research 3:201–208.

Hamblyn, C.J. 1932. Pasture establishment on reclaimed tidal flats. New Zealand Journal of Agriculture 45:78–79.

Heath, R.A. 1971. Present knowledge of the oceanic circulation and hydrology around New Zealand – 1971. Tuatara 20:125–140.

Kuchler, A.W. 1972. The mangrove in New Zealand. New Zealand Geographer 28:113–129.

Marine division, ministry of transport. 1979. New Zealand Nautical Alamanac and Tide Tables for the Year 1980. Wellington, New Zealand, 176 pp.

Ministry of transport. 1978. A guide to reclamation procedures. Miscellaneous Publication, Ministry of Transport, Wellington, New Zealand, 36 pp.

Morton, J. 1976. Mangroves and reclamation. Information Series *No. 13,* 5 pp. New Zealand Nature Conservation Council, Wellington, New Zealand.

Morton, J. and M. Miller. 1968. The New Zealand Sea Shore. Collins, London, 653 pp.

Nature conservation council 1976. Mangroves in New Zealand. Information Series No. 12, 7 pp New Zealand Nature Conservation Council, Wellington, New Zealand.

New Zealand meteorological service 1973. Summaries of climatological observations to 1970. New Zealand Meteorological Service, Miscellaneous Publication 143, 77 pp.

Ritchie, L.D. 1976. Fish and fisheries aspects of mangrove wetlands. In Proceedings of Symposium: Why are Mangroves Important? Whangarei, June 1976. Nature Conservation Council, Wellington, New Zealand.

Sherwood, F.G. 1963. Reclamation of tidal mudflats in Hokianga County. New Zealand Journal of Agriculture 107:122–125.

Walsh, G.E. 1977. Exploitation of mangal. pp 347–362 in Ecosystems of the World: I Wet Coastal Ecosystems. Ed. Chapman V.J. Elsevier.

Walsh, R.S. 1967. Nectar and pollen sources of New Zealand. National Beekeepers Association of New Zealand Incorporated Publication. 56 pp.

Wells, A.G. 1980. Distribution of Mangrove Species in Australia. Abstracts, 2nd International Symposium on the Biology and Management of Mangroves and Tropical Shallow Water Communities, Port Moresby, Papua-New Guinea, 21–55 July 1980.